The 12 Principles of Manufacturing Excellence

A Leader's Guide to Achieving and Sustaining Excellence

The 12 Principles of Manufacturing Excellence

A Leader's Guide to Achieving and Sustaining Excellence

LARRY E. FAST

CRC Press is an imprint of the
Taylor & Francis Group, an **informa** business
A PRODUCTIVITY PRESS BOOK

CRC Press
Taylor & Francis Group
6000 Broken Sound Parkway NW, Suite 300
Boca Raton, FL 33487-2742

© 2012 by Larry E. Fast
CRC Press is an imprint of Taylor & Francis Group, an Informa business

No claim to original U.S. Government works

Printed in the United States of America on acid-free paper
Version Date: 20110603

International Standard Book Number: 978-1-4398-7604-6 (Hardback)

This book contains information obtained from authentic and highly regarded sources. Reasonable efforts have been made to publish reliable data and information, but the author and publisher cannot assume responsibility for the validity of all materials or the consequences of their use. The authors and publishers have attempted to trace the copyright holders of all material reproduced in this publication and apologize to copyright holders if permission to publish in this form has not been obtained. If any copyright material has not been acknowledged please write and let us know so we may rectify in any future reprint.

Except as permitted under U.S. Copyright Law, no part of this book may be reprinted, reproduced, transmitted, or utilized in any form by any electronic, mechanical, or other means, now known or hereafter invented, including photocopying, microfilming, and recording, or in any information storage or retrieval system, without written permission from the publishers.

For permission to photocopy or use material electronically from this work, please access www.copyright.com (http://www.copyright.com/) or contact the Copyright Clearance Center, Inc. (CCC), 222 Rosewood Drive, Danvers, MA 01923, 978-750-8400. CCC is a not-for-profit organization that provides licenses and registration for a variety of users. For organizations that have been granted a photocopy license by the CCC, a separate system of payment has been arranged.

Trademark Notice: Product or corporate names may be trademarks or registered trademarks, and are used only for identification and explanation without intent to infringe.

Library of Congress Cataloging-in-Publication Data

Fast, Larry E.
The 12 principles of manufacturing excellence : a leader's guide to achieving and sustaining excellence / Larry E. Fast.
p. cm.
Includes bibliographical references and index.
ISBN 978-1-4398-7604-6 (alk. paper)
1. Production management--Quality control. 2. Total quality management. I. Title.

TS156.F36 2012
658.5'62--dc23 2011017103

Visit the Taylor & Francis Web site at
http://www.taylorandfrancis.com

and the CRC Press Web site at
http://www.crcpress.com

To my parents, Marge and "Rebel," and to my children, Jennifer and Scott, who have been the inspiration for most of the good things I've done in my life.

Contents

SECTION 2 Leading Manufacturing Excellence

SECTION 3 Appendices*

* Appendices B (Manufacturing Excellence Audit), E (Example of a Communications Plan Calendar 20XX), and F (Example of a Training Plan) are in the accompanying CD.

Preface

My leadership experience really started in fourth grade when I was elected class president. That was the first time I realized that others looked to me to provide direction, to organize and lead the implementation of important work, and to show the initiative and the passion to make things better. That first project was to plan a fund-raiser for the CARE organization. (That's where the expression "sending a care package" came from.) Our mission was to help starving children in Africa. My recollection is that we raised the handsome sum of $25.00. It was 1957.

Over the next 50 years I would be called on to lead in many other venues: more elected offices in school; chair of various committees for the Student Foundation or the Fraternity; board member and president of not-for-profit organizations such as Toastmasters International and Junior Achievement; commissioned officer in the U.S. Air Force; coach of my daughter's and son's Little League baseball teams; member of advisory boards at three different universities; various supervisory and management positions in the wire and cable industry for 10 years before devoting the next 25 years in the industry as a senior executive for two different wire and cable companies.

Over this period of time I had the good fortune to work for some excellent mentors and some very supportive senior management teams. You can learn a lot by just being in the same conference room or on the same airplane with other good leaders. You can also learn a lot from those in other walks of life who also participate on advisory and nonprofit boards and in leadership seminars. And you can learn a lot from the best and brightest on your staffs as well. I know that I have. Being observant is an important trait for anyone who aspires to leadership roles. After all, aren't we all just a composite of the people and styles we've experienced our entire lives?

As I sit down to begin the book I've been thinking about for the last 15 years, I just had to reflect on some of the people who have helped shape me and have taught me how to really think about my profession as a manufacturing leader. Each of these family members, former teachers, coaches, and coworkers stands out in my mind as an important influence on development of the style that has helped me be successful. (I know that doing this puts me at risk of forgetting someone who has been especially helpful in my development. I can only hope that if I forget to name someone that I'll be forgiven for the inadvertent oversight.)

Clarence Fast—My father, "Rebel" as he was known, who showed me by his example every day the importance of hard work and inner drive to get the most out of limited God-given abilities.

Marjorie Fast—My mother, who taught me persistence and tenacity to achieve my goals.

Mrs. Moore—My fourth grade teacher, who gave me the encouragement and the confidence to step to the front.

Duane "Cricket" Lee—My Little League baseball coach, who taught me about anticipation and about being in the right place at the right time—without being told.

Mr. Nesbitt—My eighth grade track coach, who taught me to find alternative ways for success. (When you aren't quite fast enough for the dashes, take up broad-jumping and low hurdles instead.)

Joe McPartland—My first boss, the manager of the local IGA grocery store. As a sophomore in high school I learned a lot about customer service, working efficiently and seizing opportunities to learn new things, all for $.65 per hour.

Archie Durham—My boss when I worked summers at the mold factory where Dad was a supervisor. I learned three important lessons: (1) just because your dad works here doesn't mean you don't have to do your job and make the rates; (2) if the crew gets behind it's OK for the boss to roll up his sleeves and help out; (3) if you are the supervisor of competent workers who produce quality product with a good work pace, leave them alone; just make sure they have what they need to get the job done.

Colonel Crowley—U.S. Air Force, who taught me you can be a tough-minded person but still have a good heart and recognize important achievements. For example, he took the time to write a personal note to my parents to tell them that I had been selected the Junior Officer of the Year in 1970 at Francis E. Warren Air Force Base in Wyoming. He didn't have to do that; very few knew what he did, but he did it anyway.

Dottie Clemens—Director of Special Services, my boss at 15th Air Force, March Air Force Base in California, who taught me the importance of great planning and discipline.

John Consolino and **Don Platt**—My first boss at Belden Wire & Cable (John) and the personnel manager (Don), who brought me into the company and into their management training program right out of the military. These were two seasoned veterans who knew the business well, took the importance of employee relations very seriously, and taught me the significance of involving associates in decisions that affect them in a factory. They also emphasized to me that management should be visible and accessible. They sought to have a personal relationship with everyone in a factory that employed 1,600 people when I started there in 1972.

Lloyd Shaw—My boss, who trusted me to navigate the plant through the severe material allocations and shortages caused by the oil shock in 1973–1974. He then demonstrated complete confidence and support in my guiding the plant through the most severe capacity changes in the 50-year history of the plant as we rode out a severe recession in 1975–1976.

Roland Miracle—The first plant manager to whom I reported, who taught me a lot about mentoring by exposing me to the thinking behind important processes and then involving me in the implementation to cement the learning. For example, he instilled in me the science of job evaluation and how the Hay System of salary administration works. He also communicated the important linkages of the human resource systems such as position descriptions, writing meaningful personal objectives, development plans, making succession plans, and creating and administering salary plans. He became my model of how to mentor high-potential folks who have worked for me over the years.

Nish Teshoian—The president of Belden Wire & Cable, who coached me on the importance of focus on the significant few and a no-excuses accountability for performance. He also taught me how to close a factory the right way—that is, make the tough-minded business decision to close the plant with your head but implement it with your heart relative to the people and community issues. Unfortunately, these are skills I had to use far too often in my career.

Greg Kenny—President and chief executive officer of General Cable Corporation, who actively demonstrated his complete support for my manufacturing excellence initiative during a time when I was a lone voice in the wilderness. He gave me full latitude in causing a manufacturing revolution in the company, knowing that it would be actively resisted early on by significant numbers of long-time plant managers and staffs who did not want to change. He made it clear that this would not become a "flavor of the month" and that folks in the field and the office could get onboard or else were welcome to work somewhere else. This strong support led to having several plants soar to numerous awards for manufacturing excellence and the rest of the plants to greatly improve by taking better care of customers, shareholders, and associates. It ultimately was the root of a major culture change company-wide as the support for our initiative grew throughout the company. Greg was simply the best boss I ever had.

Finally, there are countless plant managers and staff managers who have had a positive effect on my thinking over the years. Most of them now share much of my same thinking around manufacturing excellence, and all have been successful in their chosen fields. Some have stayed in wire and cable. Others have stayed in manufacturing in different industries. Still others have excelled in completely new fields such as education, private enterprise, and service businesses. This is where the biggest risk is that I'll forget important contributors, but here's my best attempt to recognize the especially influential and helpful people along the way:

> Joe Antal, Ray Bellinger, Betty Blunk, Frank Brown, Joe Consolino, Bill Darby, Dick Frist, Ray Funke, Bill Garibay, Ray Helton, Doug Hitchon, Dan Jessop, Mike Kelley, Dick Kirschner, Cindy LaBoiteaux, Jack Lawless, Lisa Lawson, Kathy Lucid, Dan Marascalchi, Joyce McDonald, Ron McNew, Steve Messinger, Mike Monti, Lee Moorman, Mike Murphy, Harry Nelson, Jim Norrick, Peter Olmsted, Jim Page, Jim Porter, Ernie Reynolds, Luis Rosete, Harry Ryan, Mike Sgro, Tedd Simmons, Bob "Cousin" Speed, Glen Starbuck, Tim Tanner, Mark Thackeray, Bob Vokurka, Lou Weber, Rick Wells, Bill Wilson, Buck Wright, Bill Yankovich, Art Yaroch, German Zavala, Jerry Zurovchak.

Again, I apologize to those I may have forgotten at the time of this writing, but thank you so much to all those who have tolerated my persistence, tenacity, passion, unreasonable expectations at times, and all my other frailties. We've all learned a lot together and accomplished important results for all those customers, shareholders, and associates who have depended on our leadership and performance to keep the big wheels turning.

This book is written on behalf of all of the aforementioned people as well as the thousands of hourly associates and hundreds of salaried staff over the years who bought into the vision, used the toolset, and have or are now executing their quest for manufacturing excellence.

Finally, I'd like to provide special recognition and thanks to Greg Kenny, Mark Thackeray, Bob Siverd, Lisa Lawson, and the incredible team at General Cable Corporation. They have provided generous support of this endeavor from the very beginning. They have also allowed me to use pictures, audits, and other forms developed during my time with them. This will make the book especially useful to the hands-on practitioners of these processes. Without General Cable's help this book would not have been possible in its current form.

Larry E. Fast

Introduction

This, my first book, is intended to share my learning from a 35-year career in manufacturing leadership positions. My hope is that the experience and the thinking behind these words will help others who share my passion for excellence and that their learning curves will be accelerated. If just one reader is inspired to step up and lead the manufacturing revolution in his company then this labor of love will have been worth it. If this book causes one experienced leader to gel his thinking around a manufacturing excellence strategy and to pursue it with passion, confidence, and urgency then I will be very gratified.

Here are a few points I'd like to make up front for clarification:

- For the convenience of both the reader and the writer, the male pronoun will always be used to avoid the clumsiness of the "he/she" and "his/her" kind of thing. Please know that I have promoted and worked with some excellent female plant managers throughout my career.
- The words *associates* and *employees*, are used interchangeably throughout the book, as are the terms *purchasing* and *sourcing*.
- Since pictures used in the text of this book are shown in black and white, you are invited to refer to the CD inside the back cover of the book. Therein are copies of the pictures in color so that you can see the power of the visual management being used in these examples.
- You'll also find on the CD three complete Excel files: a sample Manufacturing Excellence Audit, a sample Communications Plan, and a sample Training Plan. I encourage you to customize these for your own use.

Best wishes for your success in achieving and sustaining manufacturing excellence.

Larry E. Fast

The Author

Larry E. Fast is a veteran of 35 years in the wire and cable industry and held senior management roles for the last 27 of those years. At Belden, where he spent his first 25 years, he was one of the youngest plant managers to ever take the helm at their flagship plant in Richmond, Indiana. At that time the plant employed over 1,200 people in a building of nearly 800,000 square feet. In 1982 he became the senior manufacturing leader of the Electronic Division, a position that he held for 12 years. During that time he conceived and implemented a strategy for manufacturing excellence that substantially improved manufacturing quality, service, and cost. He is regarded by some as "the father of manufacturing cells" in bulk cable operations in the industry. Prior to 1987, the only known application of cell technology had been in assembly operations such as cord sets and harnesses. He later started up a new cord set division for Belden and served as the general manager for 4 years. This experience helped to round him into a stronger manufacturing leader, where his passion for excellence continued with a strong customer bias.

In 1997, he joined General Cable Corporation to lead North American operations as a member of the corporate leadership team. At that time General Cable was known more as a "marketing company" that was frequently handicapped by a grossly underperforming group of manufacturing plants. After 2 years learning his new company's culture, people, product groups, and systems, he was named senior vice president of operations. After a 1999 acquisition 28 plants as well as corporate sourcing, quality, manufacturing systems, and advanced manufacturing engineering reported to him. Later, as plants were consolidated to less than 20, he was given expanded responsibility for the North American supply chain. This included the addition of supply chain planning and logistics and three regional distribution centers in addition to the 18 manufacturing facilities in the United States, Canada, and Mexico. These plants produce a diverse range of energy, communications, industrial, and specialty wire and cable products.

His vision solidified with a strategy for manufacturing excellence that was embraced by General Cable's leadership team and board of directors in 1999. By 2001 the first General Cable plant (Malvern, Arkansas) won Top 25 recognition as an *Industry Week* magazine's finalist for Best Plants in North America. By 2009, General Cable manufacturing plants had been recognized for 21 awards honoring nine plants, six of which were named winners of the 10 Best Plants in North America. Fast's plants were the first to achieve three Top 25 awards in consecutive years and only the second company to have twice had three Top 25 winners the same year in the 20-year history of the award. As evidence of the sustainability of the process, in 2009 there were three General Cable plants in the Top 20 for the third time. Also, the very first winner, in Altoona, Pennsylvania, in 2003, came back in their first year of eligibility to compete and win for a second time. The six Best Plant winners represent the successful execution of the strategy in various cultures, union and nonunion plants, and all three countries. The winners are as follows: Altoona, Pennsylvania, in 2003; Moose Jaw, Saskatchewan, Canada, in 2005; Tetla, Tlaxcala, Mexico, in 2006; and Indianapolis, Indiana, in 2007. Manchester, New Hampshire, joined this distinguished list in 2008. Altoona, Pennsylvania, won for the second time in 2009, and Piedras Negras, Mexico, joined the list of winners after having been a finalist in two prior attempts. The 2010 Best Plants competition resulted in two plants being named to the Top 20 finalists list. They are Franklin, Massachusetts, for the second time and Lawrenceburg, Kentucky, for the first time. In December 2010, the Franklin, Massachusetts, plant was named the seventh General Cable plant to be named one of the 10 Best Plants in North America—more evidence of the sustainability of the strategy and the execution of it by the current team at General Cable Corporation.

Fast holds a bachelor of science in management and administration from Indiana University and is a graduate from Earlham College's Institute for Executive Growth. He also completed the 13-week Program for Management Development at the Harvard University School of Business in 1986.

Fast is a long-time member of the Association for Manufacturing Excellence (AME) and the Wire Association and is a former member of the American Production and Inventory Control Society and the American Society for Quality. He also has served as a Section Chair for the National Electrical Manufacturer's Association (NEMA). He has served on university advisory boards including Indiana University/Purdue at the I.U. East campus in Richmond, Indiana, and the School of Applied Sciences at Miami University in Oxford, Ohio. From 2001 to 2007, he served on the industry advisory board for the Tauber Manufacturing Institute at the University of Michigan in Ann Arbor. In 2009 he joined the board of directors of the southeast region of AME. Since his retirement he has also joined the team of judges for *Industry Week* magazine's Best Plants in North America competition.

Fast has spoken at various manufacturing excellence events such as *Industry Week's* Best Plant's Conference, Manufacturer "Live," AME Champions meeting, and the international conference on Lean manufacturing sponsored by Reliability World. He has also been published in *National Productivity Review* magazine for the turnaround story at Belden Wire & Cable and was featured in two October 2006 articles in *The Manufacturer* and in *Mexico Watch* for the outstanding track record leading change at General Cable Corporation. Fast is now founder and president of Pathways to Manufacturing Excellence, LLC, a consulting company based in Gainesville, Georgia.

Section 1

The 12 Principles of Manufacturing Excellence

1

The Manufacturing Excellence Strategy

Maybe it's just the way I'm wired, but I simply can't imagine a plant manager, director, or vice president of manufacturing thinking, "My goal is to run a mediocre factory where the safety record is poor, product quality is suspect, and scrap is too high; where shop floor housekeeping looks like a landfill, equipment is unreliable, profitability is intermittent; inventory is too high, yet customers complain regularly about poor delivery performance and high prices." That said, lots of plants in this country look and perform like that is their objective. I'd much rather imagine that each manufacturing leader is passionate about giving great service to customers, earning a good return for the shareholders while engaging associates, and providing a good living wage and long-term job security for local workers.

Achieving and then sustaining manufacturing excellence should be the goal of every manufacturing leader, period. It's as simple and as difficult as that. If that's not your objective, then you're doing a disservice to all of your major stakeholders. If that's not your goal, then get out. Go do something else. In U.S. manufacturing we don't have time to waste trying to cajole or convince the leadership that they must lead their plants to sustainable excellence. As the old saying goes, "Lead, follow, or get out of the way!"

We also don't have time for the chief executive officers (CEOs) and chief operating officers (COOs) of manufacturing companies to hem and haw about their expectations for excellence in manufacturing and throughout the enterprise for that matter. It's also not a flavor of the month. It's continuous improvement. It's forever. It is the sole operating strategy for the business.

Of course it isn't easy, nor is it without some short-term cost—typically in training and in maintenance. But if you have the right manufacturing leadership in place then they must have your unconditional support and persistence to achieve excellence. Weak-in-the-knees CEOs and COOs need not apply. It's hard work. It takes years. It typically takes significant management changes. It takes extraordinary patience. It takes constancy of purpose and focus for much longer than the next quarterly report to Wall Street. Have I said it takes years? It also takes continuity, so don't forget to get your board of directors signed up early on.

I got my first multiplant responsibilities in 1982. The first few years, absent any real mentor, I struggled to figure out my role in overseeing several plants without trying to micromanage each one of them. I'm sure some plant managers out there will vouch for this annoyance. But in fall 1986, that began to change.

As a "high-potential" manager for Cooper Industries (at that time the parent company of Belden Wire & Cable) I was afforded the opportunity to attend the Program for Management Development (PMD) at Harvard University. That intensive 13-week experience, including exposure to some of the top operations' minds in the country at the time, opened up my thinking about how to lead in a multidimensional manufacturing environment. The trigger for my epiphany was a *Harvard Business Review* (*HBR*) article (Reprint #85117) authored by Steven Wheelwright and Robert H. Hayes titled "Competing through Manufacturing." (This article was adapted from their book *Restoring Our Competitive Edge: Competing through Manufacturing*, Wiley, 1984.) The basic thinking behind the article is as follows:

> Manufacturing companies, particularly in the United States, are today facing intensified competition. For many, it is a case of simple survival. For a long time, however, many of these companies have systematically neglected their manufacturing organizations. The attitudes, expectations, and traditions that have developed over time in and around that function will be difficult to change. Companies cannot atone for years of neglect simply by throwing large chunks of investment dollars at the problem. Indeed, it normally takes several years of disciplined effort to transform manufacturing weakness into strength. In fact, it can take several years for a company to break the habit of 'working around' the limitations of a manufacturing operation and to look on it as a source of competitive advantage. (p. 2)

Wow, talk about a brain stretch for a young manufacturing executive. I had grown accustomed to just "doing the best you can" to withstand the almost constant onslaught of criticism from other functional leaders. This chorus was typically led by the sales and marketing executives who constantly griped about lead times that were too long, costs that were too high, or delivery performance that wasn't good enough. Of course the division controller normally jumped in to make it a three-part harmony. Occasionally the engineering leader would chime in as well—oblivious, of course, to the high percentage of our problems on the shop floor that had been sent there by his organization. Also, the human resources (HR) executive would dutifully display his concerns (properly I might add) about the amount of overtime the hourly folks were putting in and that they needed a break. It seemed that manufacturing was always on the defensive.

In our company, the sales and marketing department seemed to dominate strategic discussions regardless of how much of the existing dysfunction was its doing, such as selling custom "short orders" that were difficult to manufacture, requiring special raw material purchases, and causing excessive scrap and downtime on critical equipment. It wanted to grow the custom business at a time when the distribution business was sold out and had lead times of several months on the highest margin standard products in the business. But manufacturing had no seat at the strategic table in those days—and this was in a manufacturing company.

Let me digress for a moment with a story from when I was the manufacturing manager in Belden's Richmond, Indiana, plant. I recall my first customer visit. It was the late 1970s. I was so excited as this was an opportunity very few people in manufacturing ever got—except, of course, when a major mistake had been made and a manufacturing person had to go take the heat directly from the customer while the salesperson sat there with an equally furrowed brow. Bill Wilmot, one of Belden's leading salespeople, was my host and we were going to see some customers in Ft. Wayne, Indiana, who had no current issue with us. It was a friendly visit, and Bill hoped to increase our market share. So I was pumped. I was going to have the chance to see one of our very best salespeople in action and to meet a real customer. But while driving up U.S. 27, Bill said these words that have been forever tattooed on my brain: "We have a great brand name and a great catalog of products that is the envy of the industry. If only we didn't have to make the stuff." Talk about knocking the air out of a conscientious and eager-to-please manufacturing guy.

That's not to say that we weren't doing OK relative to our competition. Our performance on delivery, lead time, and cost were on par with our best U.S. competitors and better than most in a highly fragmented industry. Our profits were an industry best. But that was faint praise. All it meant, as I came to understand later, was that none of us was very good—my company was just not quite as bad as most of our competitors. But we could be a lot better. We just didn't understand that at the time.*

So, armed with this kind of personal history, it was truly a game changer when I headed off to attend the 1986 Harvard Program for Management Development (PMD). For the first time in my career I was challenged to imagine that the performance of the manufacturing organization could become the reason that the company would get sales growth, radically improved service, and profits and that the manufacturing leader might actually earn a seat at the strategic planning table and influence strategic direction as opposed to simply implementing what previously amounted to a sales support plan. This was truly my eureka moment.

The next significant breakthrough in my thinking came from the *HBR* article discussed above. It suggested that every manufacturing plant's performance "can be viewed as stages of development along a continuum. At one extreme, production can offer little contribution to a company's market success; at the other, it provides a major source of competitive advantage" (p. 3). Wheelwright and Hayes went on to suggest that manufacturing's role could be thought about in four stages that are not mutually exclusive and that it is unlikely that a company can skip a stage as described below:

Stage 1: Minimize manufacturing's negative potential: "internally neutral"— manufacturing is kept flexible and reactive.

Stage 2: Achieve parity with competitors: "externally neutral"—capital investment is the primary means for catching up with competition or achieving a competitive edge.

Stage 3: Provide credible support to the business strategy: "internally supportive"—a manufacturing strategy is formulated and pursued; longer-term manufacturing developments and trends are addressed systematically.

Stage 4: Pursue a manufacturing-based competitive advantage: "externally supportive"—manufacturing is involved up front in major

* Many of us in industry did not yet realize that a new set of competitors from around the world would very soon invade our markets on many fronts. They were low-cost competitors with acceptable quality for customers. This would lead to a wave of U.S. plant closures in the coming years. Few of us were prepared for this and, to compound things further, underestimated how good these offshore competitors would get in a relatively short period of time.

strategy discussions with an equal seat at the table with sales, marketing, engineering, finance.

It took several more years for my thoughts to totally gel, but shortly after returning from Harvard my understanding of how to lead manufacturing began to take shape. We got much more customer focused, especially on our delivery performance from the plants. At that time our average "on-time" metric was mired in the low to mid-1980s. I announced the manufacturing objective called 95 × 95, which was intended to communicate that the plants would deliver customer orders on time 95% of the time by 1995. It sounded like such a bold objective at the time, but another one of our best sales people and a tough guy from Chicago, Bob Howicz, quickly popped my balloon by saying, "Hell, man, I'd like to see 88 by 88!"

In retrospect, it was my way of beginning to focus manufacturing more externally by tying our strategic theme to customer service. We were such an internally focused and isolated group that I thought that was a good first step, though I came to see later the lack of boldness in the 95 × 95 battle cry. That said, it was still important for the plants to drive service improvements. Manufacturing would never get any respect at the Belden senior staff table until customers were a lot happier about our performance.

While this initiative was modestly successful, it still didn't get us to the root cause of many of the systemic things that ailed us, nor did it bring about any real change in other functions understanding their roles in collaboration with manufacturing. But with an earnest quest for improvement we began to do extensive reading and self-study about manufacturing excellence and began to think differently about flow, constraint management, and the like. I later took this list to General Cable and began the education process there as well. Plant managers and their staffs were required to participate, though eventually some lower-level staff also became interested and did some selective reading based on their particular roles.*

In the late 1980s, I led a revolution on the shop floors of all of the plants in Belden's Electronic Division with a major cellularization effort. By 1992, 80% of the machines that Belden owned had been reconfigured

* Please see the Appendix C for a look at the manufacturing reading list that continued to expand over the years. The original list of required reading is noted. The rest were added up until 2007 when I retired. Many of those on the expanded list were recommended by plant managers and other manufacturing leaders. Note the breadth of the subjects on the expanded list and the learning that took place from strategy to improvement tools to culture.

from departments (e.g., plastic extrusion, cabling, braiding, packaging) into manufacturing cells that were organized around common processes with products that basically required the same machine operations. For example, a multiconductor cell had its own insulating machine—plastic or rubber extruders, cablers, a jacketing line, and a packaging line. The lines were typically balanced to support one plastic extruder that put the final jacket coating on the product. In those days we didn't know about value stream mapping, but we were beginning to understand flow and the work needed to constantly debottleneck the capacity constraints to create "free capacity" as long as possible and avoid the need for capital spending. What we came to understand later was the awesome effect that debottlenecking would have on improving customer service, cost, quality, market share, and operating margin.

Within the cells, operators became more collaborative. Their coworkers were now in the line of sight and could discuss issues and respond without a lot of supervisory intervention. Cells promoted more real-time interaction, where under the old department scheme an operator in cabling, for example, had no idea who sent him the insulated conductors from extrusion and had a week's worth of inventory coming in all at one time instead of being fed only quantities he needed for the next shift or two. It was a radically different way to work, but the operators were buying in. A grassfire of support spread throughout the workforces in plants that were enjoying the new way to work, receiving recognition for it, and generating great improvements. They started to believe that they did in fact have some control over their long-term job security via improved performance. It was an exciting time sowing early seeds that would later grow into a culture change so critical to an operation being able to sustain excellence.

The results got attention at the corporate level. Calls came in at the end of a quarter that went something like this: "What's going on in manufacturing there? We see you are setting new sales records with far less inventory and much better service levels than ever." And they were right: Manufacturing cycle times were slashed, some by a factor of 10. Work-in-process inventories were more than halved. Finished goods inventory was at historically low levels in relation to sales volumes while order fill was at record highs. The transition to manufacturing cells was the engine driving this significant change in performance.

As you can see, our thinking was evolving. We were like the alcoholic who must first admit to his addiction before any healing can take place. We were saying, "I am a Stage 1 manufacturing operation." We were still blowing off our toes about every day on one thing or another and simply did not have the consistency of performance necessary to improve our lot.

Based on the data we started to collect from the process, it became clear that the root of many of our problems in the plant was caused by an inferior first-pass yield. Our internal rejections and scrap numbers were preventing further improvement. So in addition to focusing on customer service improvements, we also had to significantly improve quality. We adopted a hybrid quality program that took the best from W. Edwards Deming, Joseph Juran, and Philip B. Crosby, each one a world-renowned quality guru.

With training and focus, the plants made excellent improvements by substantially reducing the number of times customers had to sort out quality issues from their shipments. Plant teams began to use more formal tools for corrective action starting with having good data as well as a cadre of people who possessed basic problem-solving skills such as Pareto analysis and cause-and-effect or "fishbone" diagramming. Scrap was reduced by 50% over a 3-year period.

It became evident across the business that manufacturing was beginning to change and was progressing from Stage 1 to Stage 2; that is, manufacturing was establishing an internal perception of neutrality with its best competitors. The bad news was that manufacturing was still a liability. The good news was that we were no more or less a liability than our best competition from sales, marketing, and finance perspectives. And, as Wheelwright and Hayes suggested was typical of a Stage 2 operation, we had invested heavily in creating manufacturing cells, in upgrading systems, and in replacing certain old equipment with modern, state-of-the-art machines. We spent a lot of Cooper Industries' money to solidify our position in Stage 2.

THE NEXT EPIPHANY

Let me fast-forward to the late 1990s. After a 4-year stint as a division general manager, I left Belden to take the senior manufacturing position at General Cable Corporation, one of the largest wire and cable companies in the world. It took 18 months or so to get to know the people, the systems, and "the way we do it here" culture. It was a major adjustment after having worked 25 years for the same company.

But as the fog began to lift, it was obvious that General Cable's manufacturing operations were Stage 1 with maybe two plants being Stage 2. The epiphany came when I realized that thinking about the four stages of manufacturing could also be applied to each of the important metrics associated with manufacturing. For example, a plant might do a great job delivering schedules on time but have a horrendous safety record. In this case, the plant might be a Stage 3 on service but a Stage 1 on safety. So where should the focus be to move toward overall manufacturing excellence? And how can the other functions more positively contribute to this end?

In addition, the same thinking could be applied to the very culture that existed on each shop floor—every one of which was different. For example, one plant supplying products to the communications industry was run by a management team that had stopped listening and communicating with its associates years ago. As a result, union leadership over the years had learned to play hardball; thus, nearly every change of any significance that management proposed turned into a war polarizing the workforce. Local management said yes, and union said no. Union said yes, and local management said no. As you might imagine, morale was poor and performance was poor—the worst of five plants serving the same markets. The last week before the plant was closed, I met with union leadership on site and listened to their proposal of why we should keep the plant open. They offered to implement a list of the things they had been asked to do over the previous 5 years, such as go from a three-shift, 6-day operation to a four-shift, 24/7 operation to better serve a growing market. Everything they had been asked to do in recent years was to increase the plant's flexibility to adjust production levels with operators who could be trained to run various machines and to help it become more cost competitive in what had become a commodity business with lots of good Asian competitors. Unfortunately for it and its community, the willingness of union leaders to change at the eleventh hour was too little, too late. We had already invested millions at another plant that was ready to do what it took to grow good manufacturing jobs in its community and grow market share for the company.

On the other end of the scale was a plant that was an average performer but, like the previous example, did not have a workforce with an interest in changing anything. Workers were stuck in the status quo, thinking they were far better than they were and led by paternalistic managers whose expectations weren't nearly high enough. Over the years the plant leadership had "spoiled" the shop floor associates by condoning the status quo and not wanting to rock their nonunion boat. And the associates had learned to manipulate their managers accordingly. Unfortunately, in the meantime their competitors had passed them by and put them on a path to oblivion. Tough-minded decisions that were "forced" on the local leadership by the corporate office were argued ad nauseum and, when finally implemented, were done half-heartedly. Over time we learned that the plant leadership took credit for all the good things that happened for the plant (e.g., wage and benefit improvements, capital investments) while the implementation of the tough business calls (e.g., workforce reductions, limited access to capital funding, the need for improved performance) were blamed on "the bad guys at corporate."

The last example I'll use regarding the different plant cultures was pervasive. Several plants were responsible for producing the brand names of their former owner. After the acquisitions nothing had changed. Plants were very protective of this production, and senior management had not insisted on developing flexibility to produce multiple product brands in all the plants that had the equipment complement to make them. Thus, there were unnecessary customer service issues, manufacturing inefficiencies, and workforce fluctuations. Corporate optimization was simply not a priority. This represented enormous opportunities for improvement.

This lack of operational integration was a real surprise as I came into the company. General Cable had a corporate mantra: "The power of one." It was discussed regularly in company meetings and used in all the print media. As I quickly came to understand, however, it was typically used in the context of the market; that is, we have one of the broadest product lines in the industry and warehouses strategically placed across the country, so you have to deal with only one company for all of your product needs. But in manufacturing there was no "power of one" kind of thinking. As I said, each plant had its own culture. Each plant had its own set of metrics. Each plant was very internally focused. In fairness, General Cable had grown over the years primarily through acquisition. As a result, buried within it were multiple company cultures—none of which featured any progressive manufacturing thinking. Clearly, there was a critical need to put one face on manufacturing!

THE ORIGIN OF THE 12 MANUFACTURING PRINCIPLES

While I had accumulated sufficient knowledge and experience at this point in my career, I had never been responsible for more than 10 plants. It was overwhelming to contemplate leading 28 plants to "get religion" and to understand the phrase "put one face on manufacturing" and to ask them to lead the effort accordingly. I thought it was important that I try to simplify and to clarify what I meant by *manufacturing excellence* so that everyone would be in alignment with the strategy. It was also important to give them some confidence that there was a clear road map for our journey—and that I was going to lead them there.

Finally, I wanted to take the Wheelwright–Hayes model of the four stages of manufacturing to another level of detail that we could use to better understand the current state for each of the plants on each of the 12 principles relative to the four stages of progression. (You'll see in later chapters that this thinking not only evolved for each manufacturing principle but also became the model for how to think about each key function in the business.) Armed with a functional strategy of achieving manufacturing excellence, built around the 12 manufacturing principles, I took my PowerPoint presentation to the boardroom.

THE MANUFACTURING EXCELLENCE STRATEGY

In late summer 1999, General Cable's leadership team and board of directors approved our manufacturing excellence agenda. The chair of the board at that time, Stephen Rabinowitz, suggested that I should immediately call a summit meeting of all plant managers to announce our new manufacturing excellence strategy. In September I hosted a meeting for all plant managers plus my senior functional staff leaders from corporate operations, several of whom had helped directly with developing the strategy.

The timing couldn't have been better, as in June 1999 General Cable had just completed the largest acquisition in the company's history. In fact, BICC was a U.K.-based powerhouse on energy and utility cables with sales in excess of $1 billion—slightly larger than General Cable at that time. (Of note, by 2008 General Cable's sales were $6.2 billion, and it was the fifth largest cable company in the world.)

So here we were. I was staring out at the faces of 28 plant managers. Some of the faces looked excited, which I took to mean they already understood the need for a comprehensive functional strategy to drive all of us to get a lot better. Other faces were blank, which I took to mean, "OK, I'm here to listen to what you have to say." The other group of faces looked annoyed to have been pulled out of their plants to come to corporate for a couple of days of meetings. The message I took from that group was, "Hey, my plant is doing fine so I don't need to change. This is for the other plants, not mine. Hurry up and get this over with so I can go back to work."

At one point in the meeting, without calling anyone out, I simply said to the plant managers, "Please look around the room today at your fellow plant managers. Within 2 years about a third of the faces will be different." Some would come forward a few months later and say, "It's been a good ride, but I can't/don't want to undergo this change in how we operate so I think I'll just retire/resign." Others would passively resist and end up getting fired in a year or two. A couple would openly rebel and be gone very quickly. The message very simply was this: The leadership team and the board of directors have already voted on and approved the strategy. We aren't going to vote on it anymore. We're going to "just do it," to borrow a NIKE phrase.

So we embarked on the road to manufacturing excellence with the following strategy.

The Manufacturing Excellence Strategy

We want to be the best at what is important to our customers and shareholders.

Objective

All plants will work to achieve and sustain Stage 4 manufacturing excellence.

The Path: Excellence Built on 12 Manufacturing Principles

1. Safety is the cornerstone.
2. Good housekeeping and organization are required.
3. There will be disciplined use of authorized, formal systems.
4. Preventive and predictive maintenance is required.
5. Process capability will be measured on all key processes.

6. Operators are responsible for product quality.
7. Product will be delivered on time to customers.
8. Visual management will be pervasive on the shop floor.
9. Continuous productivity improvement is our way of life.
10. A comprehensive, purposeful communication plan is in place.
11. A comprehensive, purposeful training plan is in place.
12. All associates will help, and it will be an Operator-Led Process Control (OLPC) culture.

The Tools

Lean thinking and Lean and Six Sigma improvement tools

The Culture

OLPC

Make no mistake: Lean thinking is at the root of our continuous improvement initiative. The insatiable quest to eliminate waste and change the culture is the driver behind the strategy to be the best at what is important to our customers and shareholders. One of the absolutely critical enablers is to use the formal tool set to attack waste and have some early successes while getting our people trained up to expand their understanding and use of the tools.

That said, I've chosen over the years not to identify a manufacturing excellence strategy as, for example, "We want to be a Lean company" or a "Six Sigma company" or a "Lean Six Sigma company" or whatever. My view is that Lean and Six Sigma are the formal tools used to deliver improvements. The important thing is that regardless of the flag we choose to fly we must all be seeking a continuous improvement culture such that all of the other labels we use result in a common outcome of manufacturing excellence capable of competing globally. Unfortunately, my experience is that this commonality of purpose isn't shared by many who market themselves otherwise.

My issue with using Lean, Six Sigma, Agile Manufacturing, et al. is that many of these practioners tend to be silent on the critical importance of building the infrastructure necessary, along with the culture, to be able to sustain improvement over the long term. In my opinion, that is the principal reason such a high percentage of companies (I've seen survey numbers as high as 80%) that start down the road on this journey end up in the side ditch after a few years.

Lots of people out there are experts on training associates in the use of Lean or Six Sigma tools. I'm happy to leave that great work to them. The key point here strategically is to take the leap of faith that these formal diagnostic and problem-solving tools work—I especially like the discipline of the Define, Measure, Analyze, Improve, and Control (DMAIC) process.

The culture change that we seek requires that we abandon the old ways of trial-and-error problem solving and become data driven. Culturally, we have to move from having a small group of firefighters who zoom in with the Superman "S" on their chests to save the day. What we must put in their place is a broad-based group of people who are into problem solving using formal, authorized processes and who celebrate not when the fire is put out but rather when they have installed a Poka-Yoke (i.e., mistake proofing) so that the same problem will never happen again. This is fundamental to creating a new culture of fire prevention. If we don't bring discipline to this process then we simply won't see the improvements that are necessary and the initiative will fail. Let's face it, if we don't deliver the improvements, interest will quickly wane starting with the senior leadership and the plant and the company will be onto the next thing.

At the conclusion of my "special" plant manager conference, it was clear that corporate manufacturing was committed to taking the giant leap of faith necessary to achieve excellence. So as all of the plant managers scurried back to their plants to get themselves organized and to develop their local strategies, the senior operations team back at corporate attacked two other significant loose ends.

COMMON METRICS

While some progress had been made since my arrival at General Cable in 1997, metrics still were not in good order. To put one face on manufacturing meant among other things that when I looked at a delivery performance metric (or any other) in one plant it had to mean exactly the same thing in all plants. To make a very long story short, in concert with the operations staff and a giant assist from the finance team we developed a common list of metrics that all plants would use. Second, we published a list of formulas as the standard work to be used for calculating the metrics.

This eliminated all the excuses for not understanding our key metrics across all plants. As you might expect, these metrics included safety, quality, delivery, cost/productivity, and inventory. (See Appendix D for definitions and formulas.)

This meant not that individual plants couldn't measure anything else but that all plants would measure at least the list of required standard metrics and would use the prescribed formula—no exceptions.

THE MANUFACTURING EXCELLENCE AUDIT (MEA)

The second thing the corporate operations team did was to create an objective measurement of the four stages of manufacturing excellence for each of the 12 manufacturing principles. The former subjective method of assigning the stages by perception of the operations leadership team wasn't that helpful. It merely started a debate among folks who didn't have the same point of view and no objective way to break the ties. We concluded that it would always be one person's opinion versus another's until there was a detailed way to audit. While the entire team helped to develop this, the vice presidents (VPs) of manufacturing who had direct oversight of the 28 plants had major input in the development of the MEA, as did the VPs of engineering and quality. Not only were their insights keen, but their heavy involvement also helped to create the ownership that would be required for us to put teeth into the audits and to make them a key driver of the improvement initiatives* (see sample MEA in Appendix B on the accompanying CD).

As I look back on it, I sincerely believe that the MEA process is the single most important tool to bring about the radical improvement in manufacturing performance over the ensuing 8 years prior to my retirement. It provided the structure for "eating the elephant one bite at a time." The audit also:

- Captured the current state for each principle
- Quantified the gap between current and Stage 4 performance
- Forced focus and accountability on the areas that scored poorly

* Special thanks go to Mike Monti, Mark Thackeray, Bill Wilson, Bruce Evey, and Cindy LaBoiteaux for their outstanding assistance in developing the MEA and improving it every year.

- Reinforced the need for continuous improvement
- Reinforced the need to hold the gains year to year—to sustain!
- Fostered the benchmarking of best practices with sister plants—promoted reaching out and shamelessly stealing good ideas (which to that point wasn't part of the organization's DNA)
- Provided a healthy competition among plant managers
- Provided a format for understanding and involvement by hourly associates
- Provided a vehicle for plant recognition
- Along with the 12 principles, put one face on manufacturing

It is also worth noting that the MEA process itself was subject to formal evaluation after each year's audit. The operations leadership team believed it was important that we "walk the talk" and to annually review the MEA as to how it could be better for the next year. Plant managers were asked to provide feedback on how well the audit worked, what they found to be unclear, redundant, silly, whatever. The outcome from this process was a comprehensive discussion by the leadership followed by the publishing of the changes that would be made. For example, as performance to key metrics improved, the bar was raised to keep the stretch in it. As focused priorities for the business changed, the things being measured were adjusted accordingly. A final example is that a specific measurement on progress toward OLPC was added at the end of each of the first 11 principles and became the score for Principle 12. There is no doubt in my mind that without the MEA General Cable would not have had eight plants winning more than 20 (and counting) *Industry Week* magazine's Best Plants awards between 2001 and 2010.

As you read the following chapters on each manufacturing principle, please think about what role you should play to help the plant team achieve and sustain excellence. Based on your position, you may have primary ownership of driving excellence on a particular principle, or you may have a support role for another member of the staff who has primary responsibility. I also suggest that you refer to the sample organization charts shown in Figure 1.1 and Figure 1.2. Figure 1.1 is typical of what will be found in most factories. Note that it is still organized around departments and represents what has often been termed *functional silos*. Figure 1.2, on the other hand, is frequently called a *matrix organization*. This one is the structure that formally recognizes the critical linkages necessary to achieve and sustain excellence. It also helps to

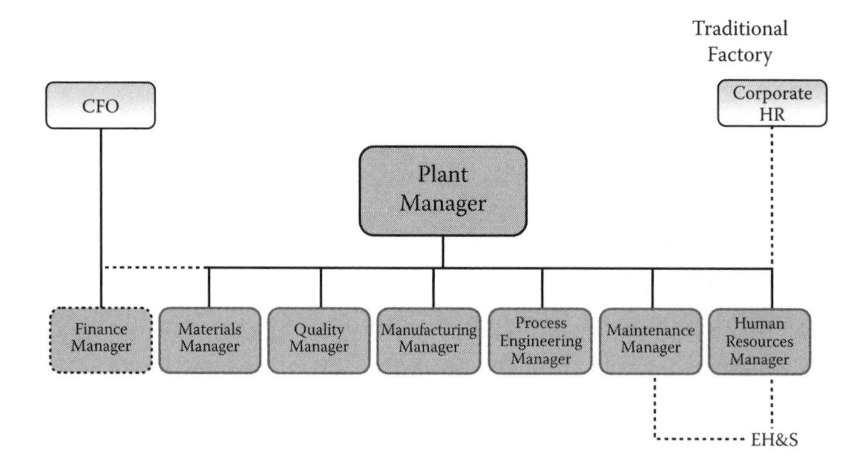

FIGURE 1.1
Sample organization chart—traditional factory. EH&S means Environmental Health and Safety. This function, if not stand-alone, typically reports to either Maintenance/Facility Engineering or Human Resources.

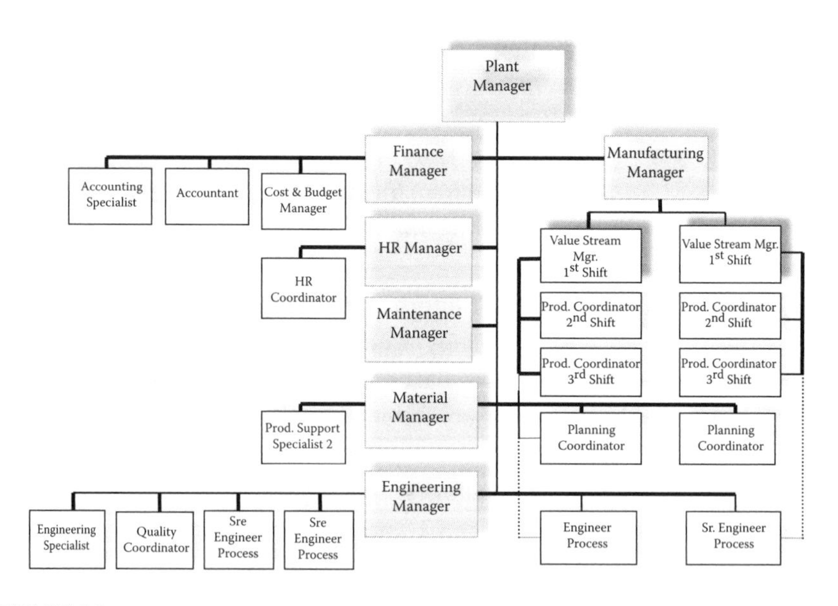

FIGURE 1.2
Sample organization chart—matrix structure.

break the paradigm of silos. In place of that, the matrix structure recognizes the importance of thought leadership and functional expertise necessary to support the shop floor and to help move the needle toward manufacturing excellence. Another way we should think about our roles is this: What can my team and I do to help eliminate all of the reasons a machine operator has a bad day?

Note that important support roles are actually documented on the matrix chart. For example, the scheduling support (i.e., planning coordinator) has dotted line accountability to the value stream manager (VSM), while the solid line report continues to the materials manager. This structure adds great clarity to the expectations of leadership. The day-to-day agenda of optimizing performance on the shop floor comes directly from the VSM to all support staff assigned to his value stream. Each support person is accountable to the VSM to execute the plan. However, the functional manager still provides the expertise to deliver functional excellence consistent with shared performance goals with the VSM. The formalization of these support relationships in the matrix often occurs with quality and maintenance personnel as well. While hourly positions often are not shown on formal organization charts, I strongly encourage the formalization of the matrix relationship with maintenance mechanics and technicians who are decentralized and matrixed as dedicated resources for the value streams. For those who would say this exposes scarce resources to being underused I would agree that the potential exists. My experience is, however, that all of the possible obstacles that may rear their heads have solutions if the correct mind-sets are answering the questions. (Hint: This includes the hourly maintenance people.)

The farther your plant progresses on the journey to excellence, the more you will see the need to transition to the matrix structure. My experience is that the change from a traditional to a matrix organization is required by the time the plant reaches Stage 3. The matrix helps to move the support staffs into the kind of cross-functional teamwork that must be in place to enable OLPC with the hourly associates.

Starting with Chapter 14, we will devote much more time to develop the new thinking by position around the staff table. But for now let's dig into more of the detailed thinking behind each of the 12 manufacturing principles in Chapters 2 through 13.

I'll close this chapter with three of my favorite quotes:

When you stop to think about it, our role in manufacturing is to continuously find better ways every day to meet impossible schedules at ridiculously low costs with perfect quality.

Larry E. Fast, former senior vice president of operations, General Cable Corporation

The reasonable man adapts himself to the world; the unreasonable one persists in trying to adapt the world to himself. Therefore, all progress depends on the unreasonable man.

George Bernard Shaw, Irish actor and author (1856–1950)

Those who say it can't be done need to get out of the way of the people who are already doing it.

Joel Barker, futurist and author

2

Manufacturing Principle 1: Safety

> Manufacturing Principle 1: Safety is the cornerstone of a high-performance plant. It is characterized by high associate awareness and involvement—effective associate training, ergonomically sound work environments, vigorous investigation, and root-cause elimination of unsafe acts and conditions—and results in zero accidents.

Why is safety the first principle? In my experience the one issue every person in the organization can agree upon is that people getting hurt on the job is a bad thing. Nobody wants to get hurt or to see a coworker get hurt. It doesn't matter if the people are union or nonunion, hourly or salaried: There is common interest in everyone being safe.

Safety performance is also a typical leading indicator of what kind of an operation the plant is. Show me a plant with a poor safety record, and I'll show you a plant that has lots of other performance issues as well. I have never seen a high-performing plant that had a poor safety record.

The plant's safety record also tends to be a direct reflection of normal working conditions. If aisles are cluttered, if maintenance is poor, if inventories are high, then all of these things tend to tie directly to higher incident rates. If a particular department or cell happens to have superior housekeeping or lower inventories, then it's likely they also have a lower incident rate for accidents. I've seen it over and over again.

Safety also reflects management's expectations. If managers are constantly walking past unsafe acts or conditions without intervention then they are speaking loudly about how they think about safety. If, on the other hand, the manager stops when he sees an unsafe act or condition and calls the supervisor of the area out to deal with it immediately, that sends an entirely different signal to the associates on the importance of working safely. I can't tell you how many times I've been on a plant tour and stopped when I saw something that required attention. Often the rest of the tour would continue on until they noticed I was no longer part of the group, and then they would come back to where I stood. I'd then ask the group, "What's wrong here? What do you see that needs attention?"

One common answer was, "Sorry, but I didn't see that when we walked past." A worse but even more common answer was, "I saw that but didn't want to interrupt the tour. I was going to call the supervisor when I got back to the office." Of course, it didn't take long for the word to get around that safety was important to me. But that wasn't the main point. I was distressed that those kinds of examples were being set every day. Folks were walking through the plant without purpose and not seeing. Others saw but lacked the initiative and the passion to intervene and to help make it a learning experience for the supervisor and the hourly associates.

This, of course, leads to my next point: Safety performance is a direct reflection of the attitudes of the workforce. For example, let's take an operator running a machine that is leaking oil now, as it has been for several weeks. He has told his supervisor about it many times, and he is aware that the supervisor canceled the scheduled maintenance to repair the leak a couple of times because his area was behind on the production schedule. Finally, the inevitable happens. The operator has been doing his best to contain the mess with absorbent "pigs," but some oil has escaped underneath it and leaked out under the machine. The operator had just replenished the supply wire on his payoff and was returning to his station to look at the control panel. As was bound to happen he slipped on the oil, fell, and tore a ligament in his knee. Now he's faced with surgery, an extended rehab, and disruption in his family's vacation plans. How do you think he feels about his supervisor right now? His company? How do you think members of his family feel about where Dad works? How do you think his coworkers feel who know about the supervisor's gross negligence?

In contrast, how do you think that operator would have felt if a different scenario had occurred? What if the first time he brought the situation to his supervisor's attention something more like this had happened:

> Ok, Brad. Thanks for bringing this to my attention. I'm going to get a mechanic over here right away to get your input so we can figure out why the machine is leaking. When we fix this I want to be sure we fix it so it won't happen again. Let me know if you guys need engineering help to get to the root of this. In the meantime, please put down some "speedy

> dry" and encircle the leaking area with pigs. That should contain the leak until maintenance can get it fixed. Use your own good judgment, but please replace the pigs when they begin to appear saturated so that the oil stays contained. The last thing we want to happen here is for someone to slip and get hurt.

Then the supervisor followed up a little later to be certain the situation had been resolved. Now how do you think this operator feels? What do you think the attitude is among his coworkers about their supervisor?

The safety record of the plant also is a direct reflection of the degree of process discipline in the operation. In the example with the oil leak, is there a preventive maintenance (PM) program? Is the equipment involved with this leak a part of it? Has the PM been successfully accomplished per the schedule? How often does equipment simply "run to failure"?

Finally, the safety performance of the plant is a direct reflection of the culture on the shop floor. For example:

- Is everyone wearing the required personal protection equipment (e.g., safety glasses, ear plugs, back support, steel-toed shoes)?
- Are operators engaged in the maintenance of their equipment (e.g., doing the necessary lubrication on a regular basis)?
- Are operators keeping their machines and work area clean (i.e., sustaining) to the prescribed standard?
- Do operators "walk on by," or do they address unsafe acts and conditions right then and there?
- Do they have the authority to speak directly with maintenance personnel and to be a part of the PM team?
- Do safety-related work orders get special priority in the maintenance department?
- Are hourly associates part of a safety team?
- Have they received proper training such as how to lift properly?
- Do they participate in accident or near-miss investigations?
- How vigorous is the accident investigation process?
- How many repeat occurrences are there?
- How comprehensive is the safety audit process?
- How thorough/meticulous is the audit follow-up?
- How do an incident and its corrective action get visibility elsewhere in the plant? In the company?

- How normal is the use of poka-yoke (mistake-proofing) as a tool?
- Do hourly associates shut down unsafe operations without being told?
- How often do you see something that doesn't belong in an aisle?
- Do hourly associates stop a coworker from an unsafe act without being told?
- Do they proactively volunteer suggestions to improve safety in the plant?
- Is a safety rule violation subject to disciplinary action?
- How frequently is safety a topic of discussion in daily huddle meetings, supervisor meetings, and plant communications meetings?
- How obvious are safe practices when an associate gives a guest a plant tour?

THE PLANT CHAMPION FOR PRINCIPLE 1

In our earlier discussion about structures of organization we acknowledged that the placement of the safety champion varies by organization. Frankly, where the responsibility resides is less important than it is to have the right leader in place who will "carry the flag" to drive improvement.

The safety champion is the person whom the rest of the organization will depend on to stay up to date with the state-of-the-art in terms of program support and training. He will be the local expert on Occupational Safety and Health Administration (OSHA) matters. He will be highly visible on the shop floor and will participate with the safety councils, safety teams, problem-solving teams, accident investigation teams, and issuance of the safety metrics. This safety zealot will use every opportunity to help shape a culture of hourly associate involvement.

Further, the safety champion will be the vocal advocate and hold up the mirror to the senior staff if he isn't seeing the kind of leadership or behavior that is necessary. He will always have the ear of the plant manager on these kinds of issues. This important staff person also understands that safety excellence is important to overall business performance and will help hourly associates to see this linkage as well. For example, lost-time accidents result in medical costs and other workers' compensation claims that must be paid. There are normally overtime costs or the use of less proficient replacement people to cover the injured associate's job. In the case of a machine operator this typically results in lower first-pass yields and lower equipment use in general. There may be negative impact on customer deliveries. Nobody wins when someone gets hurt.

Finally, I tend to think about safety the way I do about training—that is, that safety is a staff function but a line responsibility. A plant safety champion can provide great training, can conduct excellent accident investigations, and can bring in the latest programs. But if line management does not set the example minute by minute, hour by hour, day by day, then the safety results will never reach Stage 4. Every person in the plant who supervises people is a leader who must show one face on safety and be accountable. This, of course, starts with the plant manager and his direct reports.

When we get to Principle 12, we'll go into more detail about what stage 4 performance looks like in an Operator-Led Process Control (OLPC) culture. In the meantime, each of the chapters on the manufacturing principles ends with a summary showing examples of what is likely going on in the plant at each of the four stages of manufacturing excellence.

STAGE SUMMARY

Stage 1

Safety record is very poor with little operator awareness, little if any evidence of safety improvement programs, and poor follow-up, if any, to safety incidents. ORIR (OSHA Recordable Incident Rate) is > 4.0. Hourly associates are not part of the safety committee, if one exists at all. Hourly associates are not part of any accident investigations, if they are done at all. Members of management as well as hourly associates routinely walk past unsafe acts and conditions without taking any action. There is no action by management or involvement of hourly associates in developing safety procedures (i.e., job safety practices, or JSP). Safety procedures may not be a normal part of job training.

Stage 2

Safety record, operator awareness, and programs are typical of what one might find elsewhere in the corporation as well as at traditional competitors' plants. ORIR is near the corporate average and is likely in a range of from 2.1 to 3.9 and may be improving modestly year over year. A plant safety committee is organized to include hourly associates who are learning how to participate in safety improvement. Accident investigations have begun to happen, and hourly associates are being asked to participate in the process. Hourly associates are not yet proactive but are becoming more responsive to management leadership and direction. Hourly associates may be asked for input on JSPs, which are now an important management initiative for safety improvement. Boilerplate safety training has now been incorporated into job training for all positions, both hourly and salaried.

Stage 3

Safety record, safety programs, operator awareness, and involvement are better than the corporate average, with ORIR ranging between 1.0 and 2.0 and improving year over year. Hourly associates are not only active on the safety committees but also are now participating on safety improvement teams and safety audits. Hourly associates are beginning to take responsibility for their own safety and are proactively bringing attention to potential hazards for corrective action. They are also actively engaged with the writing of JSPs and with the integration of formal safety procedures into the training plans.

Stage 4

Safety record is among the best in the company and in the industry in which it participates, with ORIR of <1.0. Time between lost time accidents is measured in years or millions of hours worked. Hourly associates are well trained to perform all aspects of their job safely. They routinely teach proper safety as part of any new associate's entry into the company or into their specific work area. Hourly associates proactively seek to improve current methods and may initiate JSP reviews. They take responsibility for their coworkers' safety as well as their own by their alertness and intervention if they witness or even anticipate unsafe acts or conditions anywhere they are in the plant. Any incident is viewed as a process failure and is dealt with accordingly. Poka-yoke is a normal part of the problem-solving process.

When people work in a place that cares about them they contribute a lot more than duty.

Dennis Hayes, inventor of the PC modem

When we ponder where to begin, it becomes too late to do.

Marcus Fabius Quintillianus, Roman statesman and orator (100–35 BC)

3

Manufacturing Principle 2: Good Housekeeping and Organization

> Manufacturing Principle 2: Good housekeeping and organization are expected at all times. Use of the 5S (sort, set in order, standardize, shine, and sustain) technique is required: a place for everything and everything in its place.

Why is good housekeeping and organization the second principle? When an organization commits to making quantum leaps in performance, it's important to be able to sustain the improvement. Integral to that is a high degree of involvement by the hourly workforce in preparation of taking ownership of their jobs—in other words, in preparation for the culture change that must evolve to make long-term sustainability possible. So what is a better way to get broad-based involvement than with a 5S initiative where everyone on all shifts will have to become involved and committed?

Like safety, this principle also speaks volumes about management's expectations, commitment, and tenacity to achieve manufacturing excellence. Occasionally, a plant manager would accuse me of being too particular on shop housekeeping and organization. (I plead guilty.) My response was always the same: "If we aren't good enough to always put the broom back where it goes, then how will we ever achieve Six Sigma levels of quality on our products?"

Using 5S with hourly associate ownership is critical to establishing the attention to detail and the persistence necessary to achieve excellence. This is really about helping to create the culture that we seek, isn't it? The condition of the shop floor is the mirror we can look at every day to see how well we're doing in changing the culture in the plant. The plant manager must own Principle 2 and make certain that his direct reports are in complete alignment and are leading their teams with the same high standard of expectations. In my view, this is the first major test of how strong the leader's conviction is to direct the change necessary to achieve world-class levels of manufacturing excellence. The best I ever worked with in this regard was my plant manager in Malvern, Arkansas, Bill Garibay.

When I was on a due diligence trip in Malvern as part of the BICC (British Insulated Callender's Cables) acquisition, I was struck by the outstanding housekeeping throughout the plant. Certain processes of a wire plant, especially wire drawing where lots of oil is used, make it difficult to sustain good housekeeping. In Bill's plant you could eat your lunch right off the floor—including in the wire-drawing area.

Unfortunately, in the other plants that also had wire-drawing operations, the plant managers had fallen victim to the paradigm that it was impossible to keep that area clean. Plant managers from plants with zero wire-drawing operations also needed to see for themselves that they had no reason in their plants for having any areas with poor housekeeping. It was imperative that all plant managers were given the opportunity to see Bill's plant and to talk to him personally about his expectations and daily follow-up that led to the high standard of cleanliness.

I told Bill as we walked through his plant, "If we get this deal done, I want to have my next plant manager's conference in your plant. Several of my guys need to see this first-hand. They think I'm being unreasonable by expecting their wire-drawing areas to look like yours do every day." Of course, Bill was flattered and happy to host the conference. We got the deal done in June and had the next meeting at Bill's plant with a generous amount of time allocated for the plant tour part of the agenda. From that day forward, the good housekeeping bar was raised across North American operations. Finally, the message came through that a dirty plant was simply a reflection of the plant manager's expectations.

Also like safety, if something is out of place in the cell and any member of management walks by it, then we have just communicated volumes about our expectations and our own level of commitment. If there are too many containers in the kanban, we must stop and ask why and determine how the operator will restore order. And, yes, if the broom isn't on its hook, we have to ask why and cause correction until an hourly worker begins to see the disconnect and resolves it himself without being told.

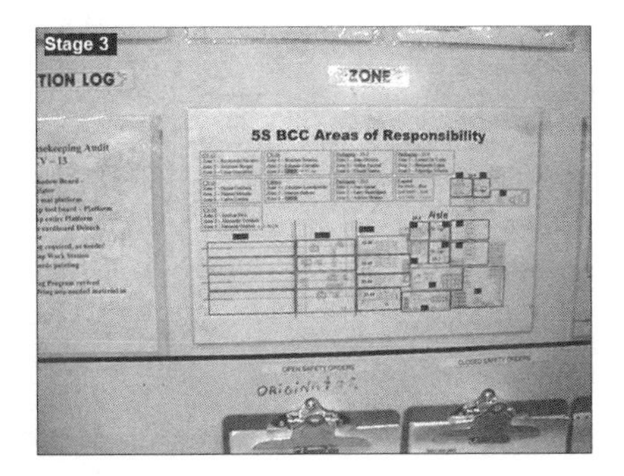

FIGURE 3.1
Stage 3: 5S common areas—ownership and accountability. (Courtesy of General Cable Corporation.)

More sophisticated visual management techniques will be used later in our evolution (see principle 8), but for now 5S is the focus. The best plants divvy up certain responsibilities on each shift such as sweeping or machine cleaning. The most important thing about 5S is that every square inch of the property, inside and out, has somebody's name on it as the one responsible for its cleanliness and organization (e.g., every aisle, every machine, every desk, every storage area, every dock, every truck apron, every office hallway and break room). Spread the burden so that every associate on every shift "owns" part of the responsibility. And use visual management. Publish the list of accountability and the schedule for each duty (e.g., daily, weekly, monthly) on the cell bulletin board. I also love to walk into a department or work cell and see an engineering drawing of the area marked up to show each area that's been designated for attention including the name of the associate who is responsible (Figure 3.1). The second most important thing is to relentlessly follow up until the new culture becomes "the way we do things here." Fortunately, training all hourly associates on 5S is not difficult. Everyone has the capacity to participate, add value, and deliver excellent results.

SORT

The first thing you should do here is to take pictures of the current state. This will be important later so you can hang them next to the improved current state after 5S and see the tremendous difference. Then have a

FIGURE 3.2
Stage 1: A place for nothing, nothing in its place. (Courtesy of General Cable Corporation.)

red-tag event: that is, place a red tag on anything that is not necessary to accomplish the mission in that area, and then remove all of the red-tagged items. These are the things that are just taking up space, collecting dust, getting in the way, and don't need to be there at all. Their removal creates space for the operator and other interested parties to collaborate and design the most efficient layout within the work area (Figure 3.2).

SET IN ORDER

Organize the workplace such that everything has a place. For example, the operators will know best where a particular wrench needs to be positioned at the point of use. Decisions will be made with the scheduling group regarding how much raw material needs to be staged at the machine and exactly where it should be placed. And, yes, the operator will decide where the broom goes (Figure 3.3).

STANDARDIZE

Now that excess material has been removed and we have established what is essential to stay in the work area, it is time to standardize within the cell and ultimately throughout the plant. This commonality will be important to everyone's ability to move seamlessly from one part of the plant to another as job opportunities present themselves—for example,

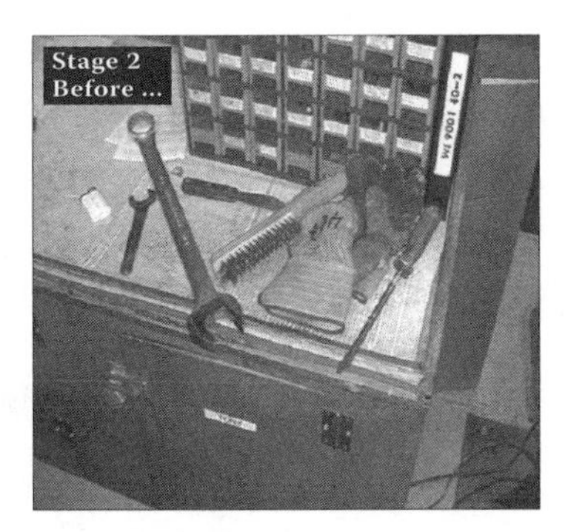

FIGURE 3.3A
Stage 2: Basic tool board—before. (Courtesy of General Cable Corporation.)

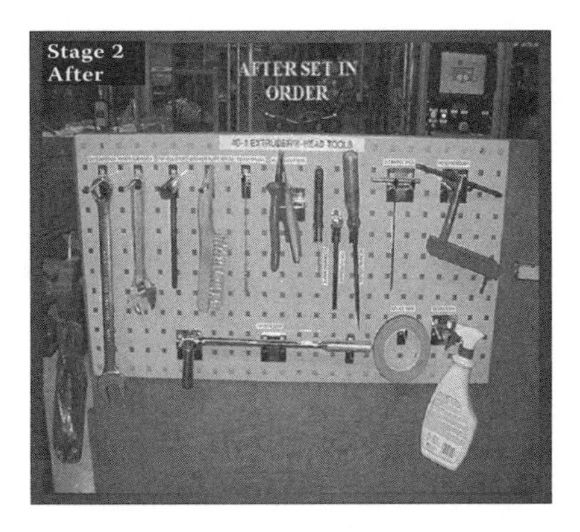

FIGURE 3.3B
Stage 2: Progress, basic tool board—after. (Courtesy of General Cable Corporation.)

a common template for showing a standard set of metrics plant-wide; a common color scheme for identifying replenishment signals for the kanban; a common purpose for Andon light colors; or a common cabinet for the storage of supplies (Figure 3.4). It is also very helpful for management folks to have a common template as they walk around the shop and "read" what's going on. The best plant I ever saw that was totally managing the shop floor visually was a Yazaki plant near Monterrey, Mexico. Any of us could have taken a tour group of strangers through that plant and it would have been clear how well the plant was performing. In my experience 5S is the best place to start to ultimately get to a shop floor that is visually managed.

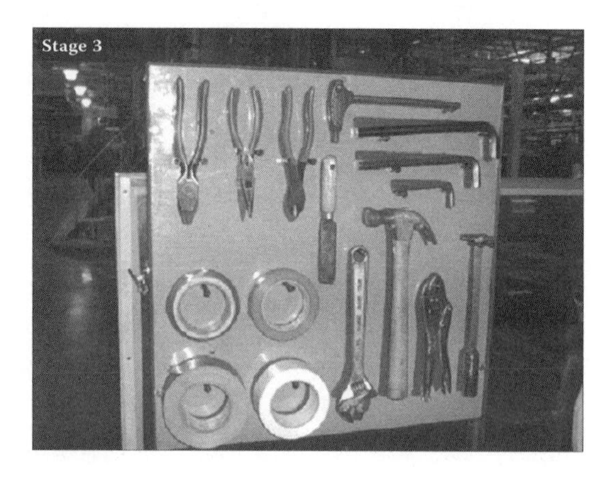

FIGURE 3.4
Stage 3: Shadow board for point of use tooling. (Courtesy of General Cable Corporation.)

SHINE

Now that everything is in its place, it is time to clean and shine. At the end of the shine phase I always like to take snapshots of how the area looks to demonstrate how it should look from that day forward. Then post the photos on the bulletin board next to the "Before 5S" pictures you took the first day. These "After 5S" pictures are the new standard. In the case of the delivery of a new machine, simply take a picture of its condition the day it is installed and hang it up. That's the standard. If there is ever any confusion about what the standard is for housekeeping and organization, just look at the pictures (Figure 3.5 and Figure 3.6).

FIGURE 3.5
Stage 3: Clean and shiny floors in a wire mill. (Courtesy of General Cable Corporation.)

FIGURE 3.6
Stage 4: Clean compound facility with no dust residue. (Courtesy of General Cable Corporation.)

SUSTAIN

Sustain: not for a week, a month, or a year—forever. This is the point where a common audit form should be developed so that at least once a week we're measuring how well each work area is doing. The supervisor may initially perform the audits, but they should later be rotated among the hourly associates. It should take only a few weeks until the hourly associates begin to feel comfortable and competent enough to do the audits. Of course, the supervisor by then should be routinely observing 5S as he makes his daily rounds. If he sees slippage that isn't being addressed by the cell members, he should intervene to train, reinforce the discipline expected, or whatever action is required to keep the cell on the right track. One of my early influences told me to always remember that "people respect what you inspect." 5S is a perfect application for that kind of thinking.

Also, perhaps a couple of times a year ask your teams to do some cross-cell or cross-department audits just to broaden the exposure of your people and to put a set of fresh eyes on the subject in their home cell. If the same people walk around in the same place every day, it is human nature that they will eventually become somewhat numb to their surroundings absent a very keen focus. Rotating the audits helps to refresh our people and helps them "get their good eye back" to apply in their own area.

They're also likely to pick up some new ideas after being exposed to a new area of the plant.

Sustaining is the toughest of the 5S's. Unless there is relentless oversight and follow-up, slippage is inevitable. I've always looked at this as the ultimate test of management's conviction—not just to 5S but to achieving Stage 4 manufacturing excellence. Once again, if we aren't good enough to always put the broom in its place we'll never have the level of discipline required to earn a spot among the elite manufacturers in the world.

Some of the observations I make as I walk around factories are as follows:

- What are the supervisor's expectations in this area?
- Is the cell organized?
- Is it obvious that the machine operator has had a major vote in the design regarding set in order?
- What is the level of inventory compared with the size of the kanbans?
- What happens to rejected work-in-process material?
- What happens when raw materials are left over after the run is complete?
- Are the results of the latest 5S audit posted in the work area where all can see?
- Are the names of those responsible, by area, posted on the audit as well?
- What is the frequency of the audit?
- Take a walk around the cell or department with the supervisor and say: "Will you please tell me what you see?"
- What is the associate recognition program for 5S excellence?

If you nail Principle 2, your operation will be off to a running start on creating visual management all over the shop floor. You'll also have a leg up on changing the culture from a traditional, status quo operation to one that is moving toward Operator-Led Process Control (OLPC). Let me repeat what I just said: 5S is critical to establishing the level of discipline required by the associates to make Stage 4 excellence possible. This discipline will ultimately spill over into all aspects of how they work.

Here is a look at what you may see evolving in the plant for Principle 2 during each of the four stages.

STAGE SUMMARY

Stage 1

Housekeeping and organization are poor. The entire area is dirty and unorganized with higher inventories present than are necessary. Equipment and process work areas are dirty. Tools are strewn about the machines and floor. There is no evidence of 5S or any associate ownership of the workplace. The only time any of the conditions are addressed is if it is mandated by a member of management with constant follow-up.

Stage 2

There is some evidence of housekeeping effort, but it is hit or miss with no standard work or evidence of 5S. Inventory levels haven't yet been addressed. Tools are stuffed in drawer or box. Management may have approached the subject of 5S or even had occasional training for sort, set in order, shine, and perhaps standardize. However, it is management directed, and there is insufficient understanding and ownership by hourly associates to have any chance of sustaining. Backsliding will be a regular occurrence.

Stage 3

Housekeeping is generally excellent most of the time. Floors and equipment are clean and free of grease and dirt. Inventory levels are significantly reduced and well organized. Most tools are on a shadow board but may not be at point of use. There is strong evidence that 5S is now being institutionalized, with hourly associates taking ownership of their work areas. Responsibility for common areas has been assigned. There is little variation in result from shift to shift or cell to cell. Everything has a place, and everything is in its place a good part of the time. Hourly associates are beginning to participate in regular 5S audits. The plant may still struggle with sustaining because corrective action still tends to be prompted by management intervention instead of being self-correcting within the work cell.

Stage 4

Housekeeping is outstanding. Inventories are sized and organized in kanbans or other pull system tools according to demand and flow. The 5S technique is being sustained in all work cells and in all common plant areas, inside and out. Plant floors and equipment are sparkling and maintained with no evidence of oil, grease, or dirt. Shadow boards or "cut-outs" are used extensively. Tools are cleaned and returned to their place after each use. Tools are located at point of use. Hourly associates now have taken full ownership of the process and are proactive with coworkers as required to sustain. Team members help each other as necessary to sustain on all shifts with no management intervention required. Everything has a place, and everything is always in its place.

> You can observe a lot just by looking.
>
> **Yogi Berra, Hall of Fame catcher, New York Yankees**

> The secret of success is constancy of purpose.
>
> **Benjamin Disraeli, nineteenth-century English politician**

4

Manufacturing Principle 3: Authorized Formal Systems

> Manufacturing Principle 3: Disciplined use of authorized formal systems is required to ensure data integrity of bills of materials (BOMs), routers, labor, scrap, and inventory records. Use of inaccurate data results in financial and customer service surprises and causes poor decision making.

I can't even tell you how many times I've walked into a plant for the first time and found the plant manager and his team basically running the plant on the back of an envelope. It is unbelievable how many plants have key operating systems filed in uncontrolled documents such as loose-leaf notebooks at the workstations or on Excel spreadsheets. What kind of control is that? What's the process for making sure all of the necessary pages are present in the notebook and that they are all the most current revision? How do you know if pages are missing? On Excel worksheets, what happens if the "programmer" dies or resigns without notice? How do the formal computer system and the worksheets communicate, if at all? How many databases are in place with which to run the plant? This is no way to run a railroad or a manufacturing plant.

First, if you work for any company that's more than a mom-and-pop shop, then your senior managers and board of directors have probably spent millions of dollars providing a formal, structured, and integrated system for the business. You can be sure they are taking for granted (naively, perhaps) that the entire organization has been trained to effectively use the formal system in the company and to exploit the powerful capability of an integrated system e.g., such as having only one database so everyone is using the same information to manage the business. However, my experience has been that this very often simply isn't the case. That's why Principle 3 contains the words *authorized formal systems*. The plant must be using authorized formal systems with which to manage the business. Here's another example.

Suppose the maintenance manager is excited about acquiring a new system for inventory control and preventive maintenance protocols, and, without formal authorization, he and the plant manager approve the purchase of a system. It may be a great system, but since they didn't follow corporate policy and formally submit a capital expenditure request (CER) for "an operating system" the company's chief information officer (CIO) had no chance to weigh in on the decision. Little did this maintenance manager know that corporate staff had been researching maintenance systems. In fact, with close collaboration with the vice president of manufacturing, the decision had been made to standardize maintenance system software corporate-wide at a huge discount, not to mention the company benefits of having a common system with which to plan regional inventory of high-cost spare parts, provide visibility of all inventory to all plants in the company, and allow standardization of maintenance and repair (MRO) type supplies. So even in this case, with a new and aggressive maintenance manager trying to do the right thing, he was off and running with a formal system that was not authorized.

The proliferation of unauthorized systems, whether formal or informal, whether on a computer or a piece of paper, will prevent the plant from ever being able to achieve and sustain manufacturing excellence in the long run. Trying to manage the business with informal, unauthorized systems simply cannot be tolerated by the leadership. Our role is to challenge the organization to find a way to meet its needs within the framework of authorized formal systems. Of course this requires that the resources will be available and provided in a timely manner by the CIO's staff.

Another common mistake I've seen, especially since the emergence of Lean, is that some members of leadership think that the materials management folks (i.e., sourcing and production control) no longer need formal manufacturing resource planning (MRP II) systems. Nothing could be farther from the truth. Any information, even if it's a wrong forecast (aren't they all?), is better than running totally blind. With the short lead times that are now typical, there is very little room for error with suppliers and manufacturing plants. Certain levels of common raw materials must still be maintained.

As my successor at General Cable, Mark Thackeray, says, "As long as there is variation in demand you need safety stock and as long as there is a lead time, you need capacity." And we both submit that this will be true forever unless someone invents a way for every company to have infinite capacity and instantaneous delivery of raw materials on magic carpets from anywhere in the world in seconds. It's not about being "anti-Lean," as I've heard a few finance guys suggest. It's about making sure that literal interpretations of "one-piece-flow" and the like don't replace common sense. Remember, our overriding strategy in Stage 4 manufacturing excellence is to be the best at what is important to our customers and shareholders. Make sure you don't discontinue the use of formal planning systems.

The only throwaways from the formal systems in my view are the shop floor systems. The cycle times should be too short and the visual management design so robust that there is no need for the formal shop floor control module of the computer system except to record what materials came in, were consumed in the process, and were shipped as finished goods to customers.

The first four manufacturing principles are what I call basic *infrastructure*. It's like being a building contractor. You have to build a strong foundation before you start putting up the walls. That's how we should think about manufacturing systems as well.

Here are some of the questions I always think about as I tour and investigate a plant for the first time:

- Is there a formal MRP and enterprise resource planning (ERP) system?
- If so, is it used with a robust, formal sales and operations planning (S&OP) activity?
- How long is the frozen time zone for S&OP planning purposes? Hours? Days? Weeks?
- Is the result of the S&OP planning used to forecast the financial performance of the business including the projection of cash flow requirements?
- How accountable is each participating function?
- Are BOMs and routers maintained on the formal authorized system? How often are they audited to ensure accuracy? What are the results of the audits in terms of accuracy? What is the process for correcting inaccuracies? Are BOMs and routers being used to back flush material usage?
- Do system inventory values match the accounting department's general ledger values?
- Are the BOMs flat or indented or structured in ways that are consistent with how material flows through the process?

- If BOMs are flat, does that mean that the cycle times are always measured in hours or no more than a few days (i.e., less than 7 days worst case)?
- If labor is still being reported, how many inputs are there? For scrap reporting?
- What is the plant's thinking and use of shop floor data acquisition systems? Do they coexist effectively with visual management?
- What about preventive maintenance (PM) systems? Are they integrated, "bolt-on," or independent systems?
- How disciplined are the users in each area?
- How well trained or knowledgeable are the users? How do you know?
- How involved are hourly associates in using the formal systems to do their jobs?
- Are data entered in real time by operators or batched for offline administrative input?

Since manufacturing is both the largest customer of these formal systems and, perhaps, their largest users in terms of data inputs, then it is important that the manufacturing manager weigh in with his organization on this principle. Since many plants don't have an information technology (IT) function, the finance manager often has the local responsibility for taking ownership of Principle 3, typically with strong support from engineering and production control, which have key roles in maintaining accurate databases. But make no mistake: If the manufacturing manager's expectations don't mirror the thinking suggested here, then he has only himself to blame because the infrastructure simply isn't going to be good enough and disciplined enough to support excellence on the shop floor. Ditto for the plant manager.

Here is what you might expect in your plant for each stage for this principle.

STAGE SUMMARY

Stage 1

All required Enterprise Resource Planning (ERP)/Manufacturing Resource Planning (MRPII) modules and other authorized systems are not being used universally, if at all. Data integrity is poor as evidenced by a lack of forecast integrity, month-end performance surprises, and regular inventory write-offs due to poor reporting of usage, scrap, labor, and other inventory transactions. Accuracy on BOMs and routers, if known at all, is poor and well below 93%* minimum accuracy expectations. Cycle counts likely aren't being done, but if they are the results are well below minimum expectations. Usage audits, if done at all, are unreliable. There's a good chance that no engineering change order (ECO) system exists and, if it does, is fragmented and unreliable. There is prevalent use of unauthorized, informal systems that are memory and paper oriented.

Stage 2

All required ERP/MRPII modules and other authorized systems are being used, and the appropriate transactions are being performed. Accuracy of BOM, router, and cycle count is measured and stable in a range of 93% to 97.4%. Implementation of a comprehensive ECO system is in process. Scrap and rejected material systems are in place and provide useful data to drive problem solving in the plant. No significant inventory write-offs have occurred in the last 2 years. Hourly associates may enter data or use certain outputs of the system, but their role is still very limited. Accounting is actively engaged with engineering and manufacturing to drive up BOM, router, and cycle count accuracy, and current costs are updated at least monthly to mirror actual raw material costs.

Stage 3

BOM, router, and cycle count accuracy is maintained at 97.5% to 98.9%. A formal ECO system is in place and working effectively. Current costs are being maintained monthly on everything except raw materials, which are being updated as changes occur. Queries are regularly used to mine data from formal systems to drive improvement activities. Hourly associates seek out data assistance to solve problems before resorting to brainstorming. All areas of the plant (and warehouses if there are any) are either exempt from physical inventories or have met the criteria for one quarter

* These numbers should be adjusted in a way that makes sense for your business based on a realistic assessment of where you are. The absolute minimum number to satisfy any professional materials organization is 98%.

and are accumulating consecutive quarters of compliance to gain exemption status. There is some evidence of the use of bar coding, scanning, and other electronic means of entering necessary transactions.

Stage 4

BOM, router, and cycle count accuracy is now measured in Defective Parts per Million Units (DPMU) terms with a goal of < 200 DPMU. (Note: By traditional measures, accuracy must be >99%, which is still 10,000 DPMU. My view is that this isn't good enough.) The formal ECO process is part of the financial forecast for future periods. Hourly associates proactively use formal systems data to drive corrective action and to initiate other process improvement projects. The entire plant and all raw and finished goods warehouses are exempt from physical inventories on the strength of sustaining BOM, router, and cycle count accuracy. Back-flushing of raw material usage is routine. There is pervasive use of electronic means of entering necessary transactions throughout the plant and warehouse to improve accuracy and reduce administrative costs.

> Our life is frittered away by detail—simplify, simplify.
>
> **Henry David Thoreau, American author (1817–1862)**

> The difference between failure and success is doing a thing nearly right and doing it exactly right.
>
> **Edward E. Simmons, twentieth-century inventor**

5

Manufacturing Principle 4: Preventive and Predictive Maintenance

> Manufacturing Principle 4: Preventive and predictive maintenance systems will be routinely used to plan and schedule equipment and facility maintenance. An undependable plant delivers poor customer service and disappoints shareholders.

This is the last of the 12 manufacturing principles that I group under the category of *infrastructure* because good maintenance (i.e., reliable equipment) is so critical to achieving and sustaining excellence. Plants that run machines to failure will be stuck in Stage 1 hell for eternity. I have never understood the mentality of maintenance managers who tolerated (and in some cases even led) this approach. It always takes longer and costs a lot more to deal with a breakdown. Unfortunately, the maintenance crew is typically known for its heroics in being able to jury-rig something to make the machine run. They are among the best firefighters in the plant (along with the expeditors), and they love the exhilaration and the recognition they get over the years. Many probably have Superman tights on under their work clothes. But make no mistake in this day and age: The maintenance department's function is to prevent breakdowns—plain and simple. There are no rewards or praise for having to react to breakdowns. There is absolutely no ambiguity on who owns Principle 4. If the maintenance manager does not step up, then the plant manager needs to be in the market for a new one.

To do this, formal systems must be used. The preventive maintenance (PM) system was a manual card system 40 years ago. It was not on a computer yet, but it most certainly was an authorized formal system. It is an antiquated notion now, but if properly managed it can still work very well. However, today's computer systems have so much more capability that there is simply no excuse for not having a good computerized maintenance management system (CMMS) in-house.

When we talk about preventive maintenance, the tendency is to assume we're dealing with production equipment. This is certainly true, but it isn't a complete picture. We're also talking about all of the key facility systems such as air, steam, water, electricity, and natural gas. Preventive maintenance on these systems is also critical. Based on the importance of, for example, certain systems, plant size, and cost of downtime, there may well be a need for a fair amount of system redundancy. For example, if you lose a major load center in the plant, is the system design such that a certain portion of power from other load centers can be redistributed to the "down" area? What if a major pump goes out on the water system? The design of facility systems is as important as the robust use of both preventive and predictive maintenance.

When touring with the maintenance manager, I'm always interested in seeing first-hand how he thinks about his role in leading such a critical function. Similar to engineering, his team must possess technical skill set requirements like those of journeymen. He may also need local experts on the unique kinds of equipment used in production of the company's products. Here are some sample questions that are always helpful in assessing the readiness of the maintenance team to help enable manufacturing excellence:

- Is there a robust training program to develop and backfill for local equipment "experts"?
- Is there a formal apprenticeship program for journeyman skills?
- Is there a formal PM system?
- If so, is there a database that includes the history of maintenance on production equipment and facility infrastructure?
- Has the PM system been put together based on these data?
- Is there a formal predictive maintenance plan in place as well?
- Is there a good record of spare parts inventory?
- What are the criteria for deciding (a) which parts will be inventoried in the plant, (b) which will be inventoried by a local supplier, (c) which will be special ordered based on need, and (d) which will be made in-house as opposed to purchased outside?
- Is there a regular "frequency of use" analysis at least annually to validate the inventory on hand?
- Are PMs scheduled through the schedule planning function?

- Are PM schedules the result of studying the work order history? Is there a formal process for revisiting the PM schedules when there is a run to failure or a close call during the year?
- Are predictive maintenance schedules based on equipment hours of run time or on calendar months? Do the methods make sense?
- Is there evidence of running to failure in the last year in spite of PM and predictive schedules?
- How involved are the machine operators in PMs on their equipment?
- Are maintenance mechanics actively involved in reviewing and modifying PM plans? Do they participate in decisions about which spare parts to stock?
- What are the key metrics on maintenance department effectiveness—for example, breakdown maintenance rate (BMR), PM work order compliance, work order cycle time for all orders (e.g., safety, constraint)?
- Are vibration analysis, fluids analysis, and heat sensing a normal part of preventive and predictive maintenance procedures? How frequently is this equipment typically used?
- Does the maintenance area actively embrace 5S?

As I said at the end of Chapter 2 on safety, show me a plant that has a maintenance department performing at Stage 1, and I'll show you a plant that will never do any better than Stage 2. Without equipment and facility reliability, the plant is always a major breakdown away from missing its customer and financial commitments. The plant is also that close to further eroding its credibility.

Here are some examples of what you may see for the four stages under this principle.

STAGE SUMMARY

Stage 1

There is little if any evidence of any kind of formal PM system. The PM schedule, if there is one, is poorly executed and is not a high priority. The production scheduling team has little if any visibility of the need for machine downtime, so it doesn't effectively plan for it in their machine hour scheduling. Operators have no role in PM and are frustrated by their equipment's lack of reliability—they almost always run to breakdown. There is no effective communication system to give operators visibility of scheduled maintenance. The BMR, if it is measured at all, is consistently >30%. Other maintenance metrics are weak or nonexistent. Machine operators often go to the break area while maintenance team members work on their machine and during PMs. The condition of the store room is poor. Inventory accuracy is unknown. Decisions on which parts to stock are reactionary and made by trial and error as opposed to being based on good maintenance data. Repeat breakdowns on the same machines for the same reasons are common.

Stage 2

A formal PM system exists and is being used to schedule PM as well as other planned maintenance. Metrics are in place or actively being developed to include at least BMR, PM schedule compliance, maintenance work order backlog by machine, and inventory record accuracy. The scheduling department has good visibility of PM machine hour requirements and adjusts available capacity for at least the current month. PM schedule compliance is >90%. BMR is consistently >15% but <30%. Data are created within the formal system to enable future planning and improvement. Machine operators typically stay at the machine during PMs but have little input and responsibility for the process. The maintenance department is doing a good job by traditional means.

Stage 3

Disciplined execution of the formal PM system is driving maintenance improvements. Key metrics have improvement goals documented, and actual performance is trending favorable to the improvement targets. The scheduling department routinely adjusts machine hour capacity as required to match the S&OP planning horizon on a rolling 3-month basis. PM schedule compliance is >95% year to date (YTD). BMR is consistently below 15% and approaching mid-single digits. Sufficient system data exist to fine-tune PM schedules and to begin analysis for predictive maintenance on major equipment and facility systems. Operators have formal PM system visibility, at the machine, of open maintenance work orders and the schedule for completion. The summary also includes space for the operators to proactively identify early warning signs that they "hear or see" and that should be checked out immediately to avoid a breakdown or more expensive repair later. Operators take an active role as part of the maintenance team during breakdowns, and PM maintenance associates routinely seek operator input prior to starting work. Operators may also be involved daily in routine PM functions such as minor machine adjustments and machine lubrication.

Stage 4

Breakdowns are rare on major equipment and always <5% BMR overall for the plant. PM schedule compliance is always >98% to the original schedule. PM and predictive maintenance schedules are reviewed and adjusted as appropriate but at least once per quarter based on available information in the PM database. All Stage 3 processes are now robust and are a documented factor in process reliability and overall equipment effectiveness (OEE) improvement on key work centers. Operators are proactive and have taken ownership of their equipment relative to both cleanliness and performance and interact routinely with their assigned maintenance person.

> The indomitable will to improve is the ingredient that separates the leaders from the pack.
>
> **Roy Harmon, co-author, *Reinventing the Factory***

> All things flourish where you turn your eyes.
>
> **Alexander Pope, English poet (1688–1744)**

6

Manufacturing Principle 5: Process Capability

> Manufacturing Principle 5: Process capability will be measured on all key processes with a minimum process control expectation of >4 sigma, 1.33 Cpk. Use of statistical tools leads to a reliable environment with predictable outcomes on quality, cost, and service. The ultimate objective is to achieve theoretical levels of both capacity utilization and material usage with process control approaching 6 sigma, 2.0 Cpk.

Let me begin this chapter with some additional clarification of this principle since I sometimes get questions when I conduct my workshop on the 12 Principles of Manufacturing Excellence. I'll speak first to the ">4 sigma" requirement.

This baseline is expected to be the absolute minimum expectation for all processes. That said, my experience is that before significant scarce resources, i.e., time and money, are devoted to improving the process beyond 4 sigma we need to be sure that the quality and/or cost benefits are worthy of the scarce resources that will be required. Often it just isn't a compelling priority. You'll need to decide for yourselves.

It's a different story, however, when we seek to achieve and sustain 6 sigma on three specific things:

- On theoretical levels of capacity utilization for constrained work centers.
- On theoretical levels of material usage.
- On the CTPs (Critical To Process) that are essential to delivering 6 sigma quality levels to our customers CTQs (Critical To Quality) specifications. (More on CTPs and CTQs will follow later in this chapter.)

The cost/benefit on these three issues is almost always a no-brainer for the assignment of scarce resources! Still, I'd recommend that an evaluation be considered once your process is stable and consistently delivering performance in the 5 sigma range, +/–200 DPMU. Of course, when a plant first starts the journey towards manufacturing excellence these kinds of questions are really hard for people to wrap their heads around. The following is much more typical of where the plant is a the start.

How many of you have ever heard a machine operator or a supervisor or, God forbid, even an engineer say something like, "Every run on this product is an adventure," or, to paraphrase what Forrest Gump said in the movie, "Life is like a box of chocolates. You never know what you're gonna get," or "I wish sales would stop taking orders for this; it's a scrap disaster every time." Well, the next time you hear something along those lines simply ask, "What is the process capability of the processes and materials used to produce that product?"

Unfortunately, my experience exposed many engineers over the years who not only didn't know what the process capability was but also couldn't have completed a process capability study if their lives depended on it. To add insult to injury, most of the time that meant that this engineer was being "led" by an engineering manager who didn't know how to do such a study either and had no expectations for his team to learn how and when to use this valuable tool in the plant. (You'll also likely need better data than a Stage 1 or 2 plant is likely to have.)

The end result under these conditions will be frequent griping by people in the plant because the sales group is bringing in orders that manufacturing doesn't know how to make. And, of course, with each rerun that has to be made due to first-pass yield deficiencies, the margins on the product ultimately shrink into negative territory. It is absolutely critical to ensure that the process capability on your processes is consistently within the specification limits with which you have to perform to make a good product.

In a traditional factory, design engineering associates stand on the sidelines, not understanding their direct role in specifying a product that regularly results in a shop floor cost disaster, an unhappy customer, and red numbers on the bottom line. Manufacturing team members continue to have the same outcome time after time though they have learned from experience to have an engineer standing right next to the operator on every run to baby the process along. They also station an expeditor in the area with his track shoes on ready to chase down enough materials to have a make-up run to try to salvage a delivery promise that is again at high risk. (What is the definition of insanity? Doing the same things over and over again and expecting a different outcome.) It's time for someone to step up and take ownership: the process engineering manager. Trust me: Your head will feel so much better when you stop butting it against a brick wall. If you are wondering where to start, here are a few suggestions.

SHORT-TERM ACTIONS

1. Rally your engineers from behind their desks and get them out onto the shop floor. Talk to supervisors and operators (their internal customers) and listen to their biggest process issues. Focus on those for improvement. Develop rapport with the operators on all shifts in their assigned areas of engineering responsibility.
2. If you or members of your engineering staff do not understand process capability and how to do the studies, then your first mission is to educate away the ignorance and train the necessary skill set. In other words, put a stop to the "dial-twisting" that goes on in the unenlightened factory. Dial-twisting, or the art of using experience only and without data, is my label for trial-and-error efforts at problem solving. It's the traditional firefighting mentally that often results in making the problem worse and rarely results in root cause elimination.
3. If you don't have the skills in-house to conduct the training, then hire an outside expert. You and your team simply cannot provide the kind of necessary support to manufacturing in its quest for Stage 4 manufacturing excellence until you have the skill set to do this kind of critical work.
4. While it may be premature to do any broad-based Six Sigma training at this point, it likely is the time for some of the process engineering staff to embark on formal Black Belt training. It is also a good time to begin training operators in the basics of statistical process control (SPC) so they can participate intelligently and be a part of the studies. Basic tools such as run charts, control limits, learning what normal variation is, and learning when to make or not make machine adjustments are fairly elementary but can be very helpful. I hesitate to suggest launching a formal Six Sigma initiative this soon. My experience

is that broad-based training on Lean tools gets the necessary resources in the game early on to harvest much of the low-hanging fruit. A Lean initiative is sufficient to drive the necessary culture change to one of seeking out and destroying waste, that is, a culture of continuous improvement! As the problems become more sophisticated, you will need much more sophisticated tools. At that time I highly recommend the addition of the Six Sigma tool set as an additional drawer in the Lean toolbox. The plant manager, in concert with his staff, should designate which positions in the plant must be Black Belts or Green Belts to support the Stage 4 performance that he seeks.

Let me give you a case based on my experience at both Belden and General Cable. When we first started on the journey to excellence we had so much inventory of all kinds and had such a void in consistent, reliable metrics that we really did not know very much about what our largest, most common problems were. Typically, it was several days or weeks before we even discovered that we had a problem, and by then the trail was cold. We just had to cope and recover as best we could. Because of the elapsed time we very often could not get to the root of what happened. We had to rely so heavily on what people could remember that it was usually a waste of time looking in the rearview mirror while we were making new nonconforming product that very minute. Of course, we wouldn't know that either for another week or two. It was insanity. One positive, I suppose, is that even though we were data poor, so many things were wrong that we simply needed to make more people aware of our performance issues and ask them to bend over and help us pick up the $100 bills off the shop floor while we got sufficient, believable data together to guide our ship. What we did know is that we had lots of problems with maintenance downtime, raw materials, and quality, but we didn't have a clue which was the most compelling for short-term help. Figure 6.1 and Figure 6.2 show the slides we used in our Lean training to describe the situation. Figure 6.1 shows the water level (inventory) as being so high as to hide the relative size of the rocks (problems) and to cause the trail to go cold before anyone in the shop finds the problem. Absent discriminating data, we could assume only that the rocks are of similar size. Figure 6.2 shows that with the level of water (inventory) down it becomes visually obvious much more quickly which issue should be attacked first: internal quality failures.

FIGURE 6.1
High water level (inventory) hides relative size of rocks (problems).

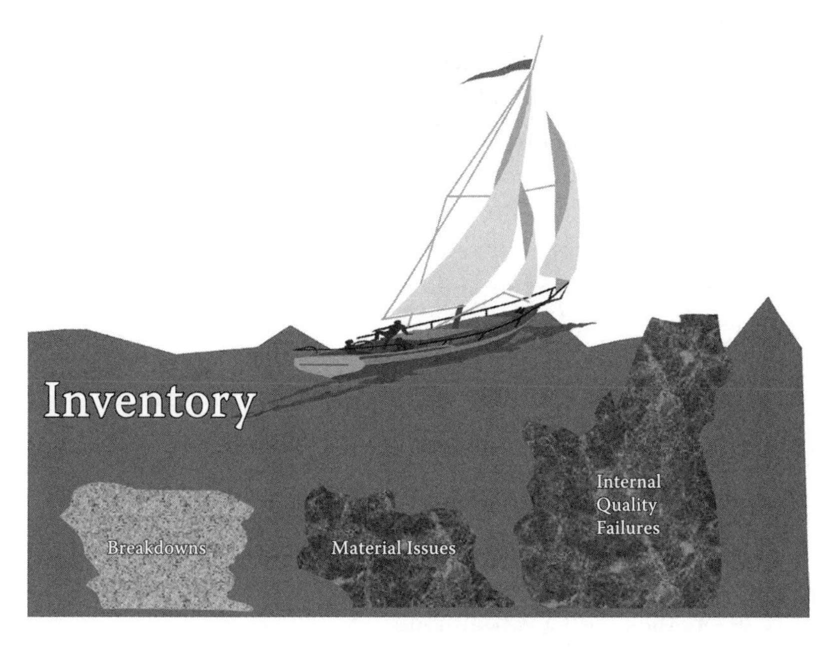

FIGURE 6.2
Low water level (inventory) makes rocks (problems) easily visible.

5. Educate and train, in addition to the engineers, manufacturing supervisors and all staff-level managers (including their boss, the plant manager, if necessary) on the concept of process capability and the need to pursue theoretical levels of performance. For example, regardless of what the current process capability is, the long-range objective is whatever the theoretical levels are. If the materials and processes being used have a theoretical/systemic scrap level of 1% and the current scrap level is 5%, then the universe of improvement available to you is 4%. This is not an uncommon theoretical universe of opportunity (TUO). Often, there is much more room for improvement than this. The same can be said about the TUO for overall equipment effectiveness (OEE) on capacity constrained equipment. What is the theoretical maximum speed you can run the extruder without compromising first-pass yield? What's the rated speed now? What is the constraint to close the gap? It's not unusual on a wire line, for instance, that an auxiliary piece of equipment is the constraint. That's like having the tail wag the dog. No thinking person would want a $20,000 machine constraining the output of a $2 million machine. An example from wire and cable that immediately comes to mind is a packaging line. That's why most small packaging has historically been done off-line.
6. Introduce yourself to the design and product engineering manager and facilitate interaction between his engineers and your own. You and he need to become best friends. Even though many companies have pursued some form of a continuous improvement culture in their operations over the last 30 years or so, I am still amazed at how frequently design and product engineers are so isolated from their process engineering counterparts—often stationed in different states.

My observation is that there is typically a lot of allegiance to the marketing folks on new product designs, but seldom has the same kind of partnership existed with manufacturing via the process engineering group. It too often is the old "throw the design (i.e., grenade) over the wall" approach. The first couple of years of a product's life cycle are typically when the sales and marketing team can command the best pricing in the marketplace—that is, before the product is commoditized.

Unfortunately, this is the time when many manufacturing operations are trying to debug the original design. It often becomes a cost-reduction bonanza for the plant at the same time that it is actually costing the company operating profit from what it could have been. This is an exercise that should be totally avoidable. How? Start with the plant engineering team giving the design team the process capability numbers on all critical processes and materials. Then if the design engineer has to wander off the proven path for a unique design (e.g., where the processing parameters are markedly different or if a nonstandard material has to be specified to meet the customer's performance spec), that becomes a warning flag to the designer to contact process engineering, to develop manufacturing trials, and to collect performance data prior to finalizing a design. The sales and marketing people then benefit from this insight and have the opportunity to participate in the debate.

They may ask questions such as, "What do you think the likelihood is that we can solve the process capability issues? Over what time frame?" Depending on those answers the sales folks may decide to take a pass on that particular order rather than risk jeopardizing the relationship with a good customer on lots of other products. Or they may think that the customer should decide—that is, is the customer willing to pay a higher price on the first few runs until we can determine our true capability (and what our real costs are) in a production environment? These kinds of interactions fall into the category of design for manufacturability (DFM) where product design engineers understand that their jobs are to send to the shop floor designs that can be manufactured within current process capabilities. This is when an engineering team truly is ready to strongly support the quest for Stage 4 manufacturing excellence. Manufacturing simply can't do it without them.

LONGER-TERM ACTIONS

I just mentioned DFM. Let me hasten to add that alphabet soups such as design for assembly (DFA) and Design for Six Sigma (DFSS) represent the same thinking and are included here. So regardless of the label your company puts on this process, design engineers must bring a few key things to the party:

1. A thorough understanding of process capability studies and the TUO thinking. The opportunity for continuous improvement is often quantified by identifying the gap between the current state and the theoretical level.
2. Information on Critical To Quality (CTQ) characteristics—ideally collected from the customer's technical people directly or, alternatively, via the salesperson. The CTQs are measurements of the finished product's form, fit, and function that are critical to their product's performance. Well-defined CTQs are what quality means to the customer. For example, in wire and cable it is often the electrical performance of the cable. However, it could also be physical dimensions on a cable that have to fit a specific connector dimension.
3. In concert with process engineering, be sure that the linkage is clear in the manufacturing specification between CTQ and Critical To Process (CTP). Let's assume that two CTQs to which manufacturing must conform are that the cable has to be smooth and round and that the cable diameter of the finished product must be to correct dimensions for the later application of a connector by the end user. The CTQs might read, "Outer jacket of the cable will be smooth and round with a diameter of .250 inches":

   CTQ = round and smooth.
   CTPs = make certain that the extrusion barrel is set at the correct temperature to melt the plastic and make it flow evenly; make certain the correct sized tooling is being used and is set up properly with the correct spacing of the tip and die; make sure the line speed is correct.
   CTQ = diameter of .250 inches.
   CTPs = gear settings, tension controls, and brake settings are correct.

Obviously, these would be much more technical and quantitative on the shop floor, but the important concept that must be understand here is that if these steps are accurately performed on each of the critical variables in the process, then the output of the process will be according to the finished product's CTQs. When CTPs are adhered to, then the product that results will be per specification. Understanding that relationship up front, in the context of process capability, will ensure far fewer surprises once the product goes into production. It is also important to remember this: Always do a measurement system analysis (i.e., a gage Repeatability and Reproducibility [R&R] study) prior to initiating a process capability study. The rationale for this is that your measurement system may have more variation than the process itself. Once you are satisfied that the gages are both accurate and repeatable then you will get a good assessment of the capability of your process. This synchronization from the customer through design engineering, process engineering, and onto the shop floor is often the difference between good suppliers and great suppliers. I'd love to hear from companies that have this degree of precision already baked into their order entry and specifications systems. I have yet to see it.

4. The process engineering manager must make sure there is a robust Engineering Change Order (ECO) process in place in the plant. It may be an authorized formal (standard) system corporately, so don't reinvent the wheel. This is a system that may be initiated by anyone as a request—typically in the form of a deviation. This is simply a warning flag that could say something along these lines: Originator is the design engineer—he reports that "the current supplier of compound X1234 plans to discontinue marketing of that product by the end of the year. This deviation is to conduct trials to replace X1234 with compound X5678 manufactured by the same supplier. The cost per pound will be the same." Or there might be a deviation that is originated by the cell supervisor who says, "On the last two production runs we had to slow down machine speed from 5,000 fpm to 3,000 fpm and have been unable to resolve the problem. This deviation is to do an analysis on the next run to determine the root cause and to provide concrete corrective action that is required. In the meantime, the current cost of the product should be based on a run speed of 3,000 fpm. This will be resolved within 30 days."

Several people receive the ECOs so that the information gets communicated well. A few, however, would have to sign off their approval of the deviation (e.g., the originator of the ECO, the process engineer, the process engineering manager, the manufacturing supervisor, manufacturing manager, the finance manager). In these days of electronic signatures there is no need for this process to cause lengthy delays.

From a control standpoint I recommend that no deviation be granted for longer than 30 days. If the problem is still not resolved at the end of 30 days, the originator will send out an update of the status along with a request to renew the deviation for another 30 days. At the end of 60 days the plant manager should be added to the sign-off list so that he can be sure the necessary resources are assigned to get the issue resolved. If at the end of 90 days a solution is not in place or within sight, then the process engineering manager will issue the formal ECO. These 30-, 60-, and 90-day milestones should be adjusted to meet your own business situation. At this time the routing, including the plant manager, is completed once again with everyone signing off on the results; for example, design engineering signs off any change to the bill of materials (BOM), process engineering signs off on any processing changes like machine speed or scrap rate assigned, and the finance manager signs off after ensuring that the current cost has been changed in the formal cost system. The key point in all of this is that a deviation is always a short-term situation that must be resolved in a matter of days.

I hope by now that the process engineering managers out there understand the huge impact you and your organization have on whether Stage 4 manufacturing excellence happens or just becomes another flavor of the month. You and the manufacturing manager and the design and product engineering manager must be joined at the hip. As the leader of the plant's technical organization, you should take the lead in establishing this kind of a relationship. It is critical to being able to sustain manufacturing excellence.

Finally, just to put into perspective the size of the challenge we often face, a long time ago I asked my quality manager to convert sigma measurements to defective parts per million units (DPMU) for all of us nontechnical types. This was done because speaking in "sigma terms" to the layperson is much harder to understand than it is to think in terms of how many units were defective per million opportunities. For the chief executive officer (CEO) to stand up and say, "At the ACME Company I want all of our products to be produced and shipped to our customers at a level of Six Sigma quality" simply doesn't say nearly as much to the average bear as it would if the CEO instead said this: "At the ACME Company I want all of our products to be produced and shipped to our customers at a quality level that results in no more than 3.4 defects out of one million parts produced."

So, on a shifted sigma to DPMU table, when we say our process capability is X sigma, then we can quickly understand how many DPMU that equates to as shown here:

Sigma	DPMU
1.0	697,700.0
2.0	308,733.0
3.0	66,803.0
4.0	6,200.0
5.0	233.0
6.0	3.4

This chart is impactful to share with the entire workforce in regular communication meetings as the manufacturing excellence initiative gains traction. It is useful to help smash long-standing paradigms where employees at all levels think they're already pretty good and don't really need all of this improvement stuff. Even a stage 2 plant operating overall at 3.0 sigma has critical processes within the plant that are operating at a much lower level than that. Another way to think about it is this: If the plant has a process that operates at 2.0 sigma, about 31% of the output (308,733/1,000,000) from that process has to have something else done to it before it can go out the door (e.g., rework, partial order remake due to scrap). If the product being made takes five processes to complete, then it is likely that only about 250,000 feet of a 1,000,000-foot order will actually go through the process uninterrupted the first time through. Isn't one of our objectives in Lean to touch things only once? Ideally, the output of every process should flow smoothly until the product is finished and delivered to the customer. Anything else is waste (i.e., *muda*).

As you walk around the shop floor and the engineering area, you can quickly determine if the things we've discussed in this chapter are the language you hear in routine conversations or not. For example:

- When you walk up to the machine do you find control charts?
- Do the machine controls provide real-time SPC information?
- Are CTP variables being tracked to control limits?
- Does the operator know the difference between normal process variation and process drift?
- Does he know when to make a machine adjustment and when to leave it alone?
- Are the engineers competent to perform gage R&R and process capability studies? Are they able to engage on the topic with machine operators?

- Are quality metrics tied directly to CTPs delivering the CTQs specified by the customer?

Here is what you might expect in your plant for each stage of this principle.

STAGE SUMMARY

Stage 1

Processes are generally not capable, and terms such as process capability, Critical to Quality, and Critical to Process are not in the vocabulary. If any data on key processes exists at all it isn't being used to drive improvements in process capability and the reduction of internal quality failures. Audits of process capability known versus the total processes available do not exist. Internal rejections typically > XXX,000 DPMUs.*

Stage 2

There may be some episodic use of process capability studies but there are few who understand or use them. CTQ/CTPs are in the early conversation stage with design engineers, process and/or quality engineers, product management, and sales. Preliminary engineering work is being done to establish upper and lower control limits. Operators are getting started with SPC by collecting and charting data to support the development of controls to prevent process drift. Internal quality rejections typically are <XXX,000 but >XX,000 DPMUs.

Stage 3

Process capability study usage is widespread and well understood by operators on all shifts, engineers, and supervisors; basic SPC training is complete, and operators are knowledgeable relative to when they make process adjustments versus where they leave the process alone due to normal variation. A process capability audit schedule has been developed and

* These should be input based on the data for your operation. I have suggested some typical dimensions for each stage based on the number of digits.

shows >95% compliance. Process remediation is well defined and under way. CTQ/CTP characteristics are prominent in shop floor documentation and are highly visible. Key processes have Cpk >1.33 and approaching 5 sigma levels of 1.5. Internal rejections are typically <X,000 and trending favorably.

Stage 4

Processes for CTQ/CTP characteristics approach Cpk of 2.0 or greater, 6 sigma levels. Operators are highly familiar with process capability and proactively use SPC to both control current processes as well as to drive further improvements. Audit schedule shows 98+% compliance. Internal failures are now at XXX and trending toward six sigma level of 3.4 DPMU.

> I cannot help feeling that people have not thought rigorously enough about the notion of improvement.
>
> **Shigeo Shingo, industrial engineer**

> There can be no Kaizen without standard work.
>
> **Taichi Ohno, co-author, *Toyota Production System***

7

Manufacturing Principle 6: Product Quality

> Manufacturing Principle 6: Operators are responsible for product quality and will not knowingly pass defective material to the next operation. The objective is to have zero escapes of poor quality product to the next operation or to an external customer.

First, let me admit to how impossible this expectation is for operators in a Stage 1 plant. They're likely the subject of disciplinary action if they have repeated cases of running material that is not per specification or if there is recurring evidence of product they ran being rejected or causing scrap downstream in the process. Sure, sometimes operators are careless and may need a wake-up call, but this is the rare exception. If you were constantly being chastised for having quality issues, though, would you continue to write the reject tags or otherwise draw attention to yourself? Human nature says we all would be more likely to let some things slide.

Most often, when a plant has high internal failure rates, management across the board has failed the operators. Until and unless their management has provided the correct processes, correct specifications, correct setup instructions, correct measurement tools, correct training, correct materials, and correct maintenance, it is not reasonable to expect near perfect quality from any operator. Fortunately, once all of these things have been put into place it is likely that, as a part of the process, the leadership has become much more enlightened.

These more enlightened leaders then develop an insatiable thirst for improvement. The what and the why become much more important than the who. And that's a major step forward in the quest for manufacturing excellence and the culture that will help to sustain it. Why are rejections high on that machine? On that product? Is it a training issue? Process capability issue? Gage R&R issue? Maintenance issue? Material issue? What can the operator tell us about what's happening? How can we find and isolate the root cause so we can solve the problem and stop producing waste? This is a great time to break out the "5 Whys" from the Lean toolbox to get to the root cause for corrective action.

And how do we convince the machine operators that we expect them to recognize quality issues and write the reject tags with the understanding that they will not be disciplined for writing them anymore? What they will be disciplined for in the future is not writing them and then having subnormal quality material from their operation show up downstream. We want to encourage putting the spotlight on problems so we can solve them, not sweep them under the rug. One of my respected colleagues over the years, Dick Kirschner, likes to say, "Make it ugly!" so all can see. In other words, make it visible and obvious so we know to deal with the situation. (I like to see the "fallout" area positioned on an aisle for easy viewing and follow-up.) Of course, it takes time for management to convince the hourly folks that they've turned over a new leaf and that they won't lose their convictions and return to Stage 1 behaviors. Management has to walk the talk every day to earn credibility with hourly associates.

QUALITY SYSTEMS EXPERTISE

As will be discussed in Chapter 19, the quality manager has to provide quality systems expertise and be the quality systems zealot for the plant. I've always thought it was critical for the quality manager to go into any certification effort (e.g., ISO 9000) with the right mind-set—that is, we are entering into this certification process with the understanding that it is just another tool to help us make the business better. This is opposed to those who get quality certifications simply to comply with customer demands or other external influences. It has to be about getting better, or it becomes a joke to the workforce and a colossal waste of time and money. Unfortunately, my experience is that external auditors on these kinds of standards often stop thinking when they come into the plant and go through the motions, to the letter of the standard, without making sure that the process being meticulously followed is accomplishing the right outcome for customers and for the plant.

For example, I've often had International Organization for Standardization (ISO) certification explained to me like this: "We document what we do, and then the ISO certification inspectors come through and audit. If we are doing exactly what we said we would do then we'll be certified." Unfortunately, we can document poor operating practices and do them exactly the same way every time all the way to both certification and oblivion. What is important if you're serious about seeking Stage 4 manufacturing excellence is first mapping the processes and making sure you have the right processes in place and then documenting them. Of course, also train all the people who need to know exactly how to execute the right processes going forward. The documentation for the certification becomes important standard work for plant control.

Getting any kind of a quality certification takes a huge investment in time and money. If your company chooses to make that investment in the business it is the responsibility of the operating team to do it the right way. And the quality manager has to be on the point.

STANDARD WORK

You may wonder why I've chosen to put this topic under Principle 6 instead of discussing it at length in Chapter 17 with the role of engineering. Certainly I could have included it there, because standard work is one of the cornerstones of engineering. On the other hand, since most plants these days are required to seek some form of quality certification, I choose to look to the quality team to champion standard work. Also, in a manufacturing plant the finished product is the source of most of the value add for the company. The notion is, therefore, that if you build in the discipline required to make near perfect products at a Six Sigma level of performance then the rest of the organization gets pulled along.

Here are some of the questions quality engineers or process engineers should ask as they seek to provide or document standard work:

- Is there a comprehensive calibration system in place for test equipment? Does it include machine gages? Does it include clear standards and procedures for ongoing and regular calibrations? Will calibration become a part of the plant's preventive maintenance (PM) program, or will it be outsourced? Are the results and the date of the last calibration audit conspicuously visible to the operator? Has the operator participated in formal gage Repeatability and Reproducibility (R&R) studies?

- Is there only one correct way to set up and start a machine?
- Is there a process for first-piece inspection to be sure the startup is correct?
- Is there a documented process for Operator Self-Inspection (OSI)?
- Is there a Poka-Yoke in place on this process?
- If not, is there a consecutive, redundant check at the next operation?
- What is the process for isolating and making disposition on any subnormal material at this workstation? What are the accounting controls on these transactions?
- What is the control on subnormal material?
- Is there a comprehensive, documented Material Review Board (MRB) process in place? How effective is it? How do you know? Are MRB materials contained in a designated area and dispositioned every day?
- If there is a rework activity, what is the input–output control? Is it centralized or decentralized, that is, controlled right at the machine or within the cell?
- What data are collected on internal failures? How are they used to drive corrective action?
- Is there a clear understanding of the direct linkage between the critical to quality (CTQ) characteristics of each product and the critical to process (CTP) characteristics of the process? Where/how is this information communicated via a formal authorized system?
- Unless the plant is performing to at least a 5 sigma level (i.e., about 200 defective parts per million units [DPMU]), is a dock audit process in place as a "last chance" to internalize any quality issues before allowing them to "escape" and be shipped to a customer?
- If so, is it statistically valid? Are the results used to drive corrective action?
- Are customer returns being measured, and are useful data being collected and used to drive correction action?
- How quickly does the sales organization get defective material samples back to the plant for analysis? What percentage of requested samples are actually returned?
- What correlation is there between the findings on customer complaints and returns and what is being checked on the dock audits?

- Are items to be checked on the dock audits ever changed based on external findings?
- Do dock audits include things like labeling and packaging in addition to the product audits?
- Is there a path for excellence that includes eliminating the long-term need for doing dock audits?

Disciplined quality systems are critical to creating the expectation of and the reality of standard work—ultimately delivering manufacturing excellence. And it has been my experience that when data on quality failures are used to drive corrective actions on the process, then the plant almost always ends up working on the right things to improve the business. By definition, solving a quality problem always results over the long term in happier customers, lower costs, better morale, and positive movement toward Stage 4 manufacturing excellence.

We'll talk more in Chapter 13 about further evolving the machine operator's role in this manufacturing principle when we discuss Operator-Led Process Control. The following section provides summaries of how you can audit your own plant relative to the four stages.

STAGE SUMMARY

Stage 1

There is evidence of poor quality being moved to next operations, to a central rework area, or to a work-in-process (WIP) storage location. Customer complaints and returns are not measured statistically and exceed 4,000 DPMU. There is little if any evidence of data being used to drive corrective actions and problem elimination. Operators for the most part are ignorant of customer issues and are uninvolved. Inspectors are checking materials and writing reject tags as required.

Stage 2

First-piece inspection is in place. OSI is in use, along with the use of some in-process controls and tests that prevent most defects from being moved to the next work operation. Scrap and DPMU metrics are in place. Defect Pareto charts, and others as appropriate, are displayed at the workstation. External failures are <4,000 DPMU. Data and customer feedback are used to implement fixes; however, long-term corrective action is not yet in place, and repeat occurrences of external defects still happen.

Stage 3

There is early evidence that operators are beginning to take ownership for the product quality they produce. DPMU, scrap, dock audit, and customer complaints and returns data are being used to do team-based problem solving. The workplace is highly visual and includes status of active quality improvement team projects. Operators take it personally if they have a quality escape from their area. Operators now have the authority to shut the process or machine down due to quality issues. External failures are <1,000 DPMU. There is robust use of data to drive sustainable corrective actions. Problems are routinely eliminated with no repeat occurrences. Poka-yoke is becoming a significant last step in problem solving.

Stage 4

Operators are proactive and into a problem-prevention mode. The use of poka-yoke is prevalent throughout the facility. Operators own quality and often solve problems at the source. Scrap is at or within 1% of process inherent or theoretical levels. External failures are at <200 DPMU and trending towards 6 sigma. Operators drive sustained corrective actions, perform follow-up audits, and perform Failure Modes and Effects Analyses (FMEAs) to validate problem elimination. Solutions are benchmarked and applied to similar processes in a preventive manner.

> We are what we repeatedly do. Excellence then, is not an act, but a habit.
>
> **Aristotle (384 BC–322 BC)**

> We must become the change we want to see.
>
> **Mahatma Gandhi (1869–1948)**

8

Manufacturing Principle 7: Delivery Performance

Manufacturing Principle 7: Product will be manufactured on time to the original, agreed-on delivery promise. Manufacturing's delivery performance is critical to growing the business profitably. It is a measure of the reliability of the shop floor and is an essential part of providing service that ultimately delights the customer.

I've always thought that this metric, more than any other, reflects how well synchronized and managed the operation is overall. Think about it: If the plant is not performing well on any of the 12 manufacturing principles, the chances of consistently delivering on time to customers are nil. I remember all the excuses from my early days as a schedule supervisor. We could have carried around a list in our pocket and just referred customers to a number for the explanation (i.e., excuse) we were giving them. "It would have been on time except…"

- There was excessive scrap and the order came up short.
- Part of the order had to be reworked so the rest will be shipped next week.
- The machine operator tripped over a pallet and broke his ankle, and I don't have anyone trained on that machine until the weekend shift.
- Two of the ten reels were lost somewhere in work-in-process inventory, and we didn't find it until it was too late to make this week's schedule.
- The final operation couldn't be run because the machine had a catastrophic bearing failure in the main drive shaft.
- The engineers couldn't figure out how to make the plastic extruder run up to rated speed.

- An operator somewhere up the line passed along defective parts, and we didn't find out about it until final test. We'll have to expedite a remake for next week.
- I couldn't possibly make this week's schedule because production control had overscheduled my 7-day capacity by 20% including all of last week's past dues.
- I didn't get this schedule delivered because I had to slow the machine down to 80% of rated speed, and I ran out of time to make the whole week's schedule.
- I found out that the production planner is scheduling my cell based on theoretical capacity instead of demonstrated capacity, so it's not my fault.
- An operator, with his supervisor's permission, overran an order because the same product is also scheduled to be run next week. This was done to avoid a setup. It was the efficient thing to do.
- The raw material came in late.
- The raw material was rejected in receiving inspection and held there for disposition. I'll try to get more material for next week.
- The operator ran the order slightly oversized, and the yield on the material came up short.
- I couldn't get anyone to work overtime to cover absenteeism.
- The prior operation sent me the material late, and I couldn't make up the lost time.
- The dog ate my schedule card.

I know anyone with an interest in manufacturing excellence can relate to all of these and can likely add many more to our list of excuses. The materials manager certainly owns Principle 7 as the "customer's advocate" for service in the plant. But it's clear that he will be a voice in the wilderness if the plant manager and the rest of the team don't stand up and say, "Enough!" on missing customer deliveries. Principle 7, more than any other, shows how integrated the various functions are in the plant and how interdependent we all are if we expect to perform at a high level. The plant must function as a well-coordinated, well-oiled machine of manufacturing excellence zealots.

As I try to assess the health of a materials and production control operation, these are the kinds of questions that come to mind:

- Is there a formal sales and operations planning (S&OP) process? Are sales, product management and marketing, engineering, and finance all involved and equally accountable for their respective outcomes at the same level as manufacturing?

- Is there a formal, integrated enterprise resource planning (ERP) and manufacturing resource planning (MRP II) system being used?
- Is the MRP module being used to plan raw material requirements?
- Are pull systems being used to execute the materials plan?
- Is capacity requirements planning (CRP) being used to ensure that the schedule plan is executable; that is, is capacity based on demonstrated?
- How timely and accurate are the data?
- What process is used to plan crew size out of the S&OP process?
- What size *band of flexibility* to meet customer demand can the shop floor manage effectively (e.g., 5%, 10%, 0%)?
- What is the *frozen time fence* (e.g., 3 hours, 3 days, 3 weeks, 3 months)?
- How is the shop floor managed (i.e., with the computer system, visually, or a combination)?
- Is the cell production schedule being pushed, pulled, or some of each?
- Is there evidence of the evolution using formal planning systems with visual execution systems?
- How well does the product flow through the plant match up with the design of the BOMs and routers?
- Is there a disciplined audit and corrective action process in place to ensure the accuracy of BOMs and routers?
- How linked is the shop floor schedule to the shipping schedule? To the customer request date?
- How much safety stock is planned to compensate for poor plant execution and for higher than forecasted demand? In the finished goods warehouse? Consigned in customers' warehouses? In work in process (WIP) on the shop floor? Raw materials held in the plant? At the suppliers' plants?
- How much extra queue time is planned to compensate for reliability issues?
- Are the reasons for missed schedules known? If so, are the data being used to help drive the improvement in the plant? If not, what is the plan to identify the reasons schedules are missed?
- Is there a formal process for selecting suppliers? For certifying them?
- Are suppliers measured to the same kinds of cost, quality, cycle time, and on-time delivery performance that we use for measuring ourselves? Is their share of the business tied directly to their performance on these key metrics?

- How involved are key suppliers in managing the raw material inventories in the plant (e.g., electronic kanbans, video-controlled replenishments, employee on site, daily milk run)?
- How proactive are the communications internally when a scheduled delivery is at risk? Who makes the contact? Customer service rep? Salesperson? Plant staff member? Machine operator?

STAGE SUMMARY*

Stage 1

The plant is consistently unreliable at scheduling and executing orders. Customer commitments are often missed despite excessive expediting. Delivery performance is typically below 80%, probably well below. The shop floor is unreliable, as are the systems that support it. A formal S&OP process is nonexistent; consequently, there is no formal method to align rough-cut capacity and supplier material availability with future loads. There is no process in place to understand schedule misses, to identify root causes, or to capture opportunity to improve performance. There is also rarely any proactive communications telling customers their delivery schedule is at risk. There are a lot of unhappy surprises.

Stage 2

The plant is inconsistent in its scheduling and execution performance to customer commitments. Proactive communications with customers aren't consistently done, though there is much more of an effort to do so. A formal S&OP process is in the early stages of development, and team members are struggling to understand their roles and to participate effectively. Rough-cut capacity and materials planning are being done, though data integrity issues are being exposed for correction. Bottleneck operations and constraints are not fully recognized, and overschedules and expedites are still the order of the day far too often. Better information is becoming available on the causes of missed schedules, but robust corrective actions are still hit-or-miss. On-time delivery performance typically is in the range of mid-80s to low 90s.

* Thanks to Mark Thackeray for his rewrite of this stage summary.

Stage 3

The plant is much more reliable at planning and executing schedules to the customer's delivery date. External commitments are typically being met at a rate in the mid-90s. The formal S&OP process has gained traction. Suppliers are provided with material projections for their planning, and the multifunctional S&OP team is working well together and making improvements in the process. Schedules are screened against rough-cut capacities, and constraints are now finitely scheduled. Natural work teams in the cell are beginning to manage the pace to Takt time and seldom require expedites. Schedules at risk or missed are reviewed daily in the cell, and a data-driven approach is used to identify root cause of misses and to initiate corrective action.

Stage 4

The plant rarely misses a customer commitment and is now a tool for the sales team to help win new business. The S&OP process is now mature and routinely delivers leveled loads to the plant consistent with materials availability and machine and labor capacities. Production output is measured against Takt time. Operators in the cell are now multiskilled and in control of their processes. Corrective actions typically include poka-yoke.

> Even if you're on the right track you'll get run over if you just sit there.
>
> **Will Rogers, American humorist (1879–1935)**

> I've never been satisfied with anything we've ever built. I've felt that dissatisfaction is the basis of progress. When we become satisfied in business we become obsolete.
>
> **Bill Marriott, Sr., former chief executive officer, Marriott Hotels**

9

Manufacturing Principle 8: Visual Management

Manufacturing Principle 8: Evidence of visual management will be prominent throughout the plant with key metrics at each work center. Kanbans, Andon lights, alarms, color coding, and other visual techniques will be used as appropriate.

Some have asked why visual management is included in my 12 manufacturing principles. There are several reasons I think it is important:

- We learn better when we can see it. We are all visual creatures.
- Visual management requires that we think hard about simplification. Eliminating waste always results in a simpler process. With Lean thinking, simple and effective are always better.
- Decisions made from visual cues and data are more real time than computer reports.
- It helps us to develop a better understanding of the process and the hourly associate's role in moving the needle on his cell's performance.
- It provides a great forum for operator involvement in contributing to the design of the visual system.
- It provides a model of the discipline that is required to achieve and then to sustain manufacturing excellence.
- It provides the means for operators to take responsibility for their work by making it immediately obvious to everyone when something requires their attention.
- It is one of the best and easiest ways to begin operator engagement.
- It is highly supportive of the development of the Operator-Led Process Control (OLPC) culture on the shop floor.

The ultimate test of a Stage 4 plant is when you can take your spouse or a friend who is unfamiliar with the operation through your plant and they can figure out what is going on with little explanation. It should be that obvious and simple. One of the best visually managed plants I ever saw was a Yazaki assembly plant near Monterrey, Mexico, in the mid-1990s. When I walked into the plant, the first thing I noticed is that nothing in the entire plant was more than chest high—even the employee lockers. You could literally see the entire plant from the entrance, and it was spotless.

As we walked through the various assembly cells it was easy to understand what was going on. There were value stream maps posted at each cell. This, of course, allowed anyone to understand the flow of the product through the cell. All of the key metrics, such as safety results, schedule delivery performance, constraint overall equipment effectiveness (OEE), scrap or any quality defects, amount of inventory authorized and actual, and housekeeping assignments, were clearly posted in each cell. There was also a communications board for plant-wide information as well as any news specific to the cell. The template used was exactly the same for all cells. The metrics were displayed as easy-to-read graphics supported by the numbers. The entire plant was highly disciplined and managed in a 5S fashion. Visual aisle markings were absolutely sacred.

Some 10 years later, I was privileged to have a communications cable plant in Tetla, Tlaxcala, Mexico, which won an *Industry Week* magazine Best Plants in North America award for 2006. The Tetla plant became the benchmark within General Cable Corporation for its robust use of visual management. This was an operation with a legacy, multibuilding campus layout that added to the complexity of the operations. But the locals' passion for manufacturing excellence stimulated them to find a way in spite of the less than ideal plant footprint. Operators could call for a forklift, raw materials, maintenance, or engineering support by simply flipping the correct switch that illuminated the proper light signal visible to anyone in the building. In one area they put sensors in the floor to trigger the replenishment of kanbans. In other areas where such sophistication wasn't necessary they simply painted the floor in stripes of green, yellow, and red based on the engineered size of the inventory area and the necessary replenishment cycle time. Andon lights made it easy to see when all was well and impossible to ignore when an operator needed help. An enhancement could be the addition of an audio signal after a certain amount of time has elapsed if the necessary support group has not yet responded.

A simple rule of thumb for applying visual management is that when there is a need for standard work, always ask the question, "How can we make this visual?" The creative thinking of hourly and salaried team members usually produces the right tool to make it happen on virtually any topic such as job instruction, cross-training schedules, kanban controls, maintenance assistance, machine speed variance, and to poka-yoke quality issues. Figures 9.1 through 9.10 are pictures of visual management at its best.

FIGURE 9.1
Stage 4: Color-coded kanban replaces paper-based work orders. (Courtesy of General Cable Corporation.)

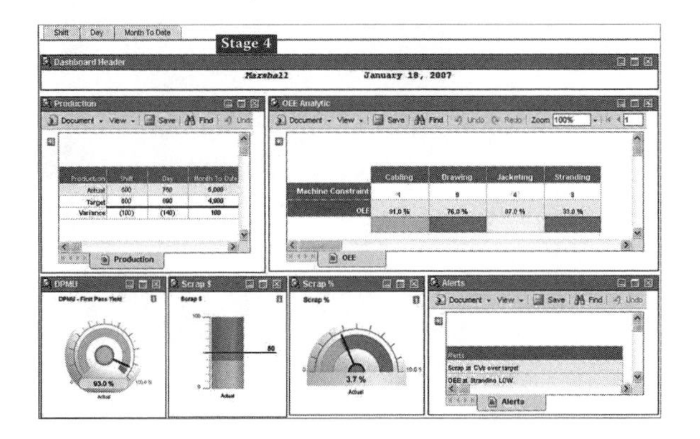

FIGURE 9.2
Stage 4: Dashboard of key performance indicators for cell associates. (Courtesy of General Cable Corporation.)

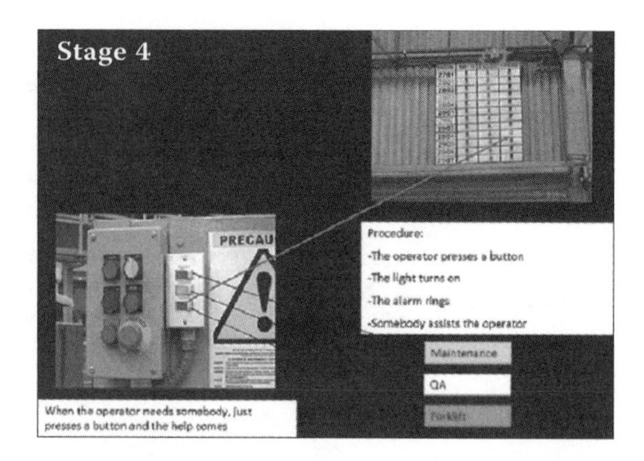

FIGURE 9.3
Stage 4: Push-button call system. (Courtesy of General Cable Corporation.)

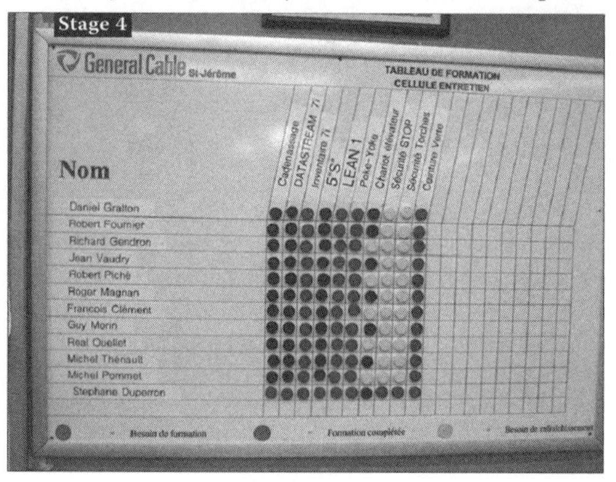

FIGURE 9.4
Stage 4: Cross-training matrix for operator skills, managed by operators. (Courtesy of General Cable Corporation.)

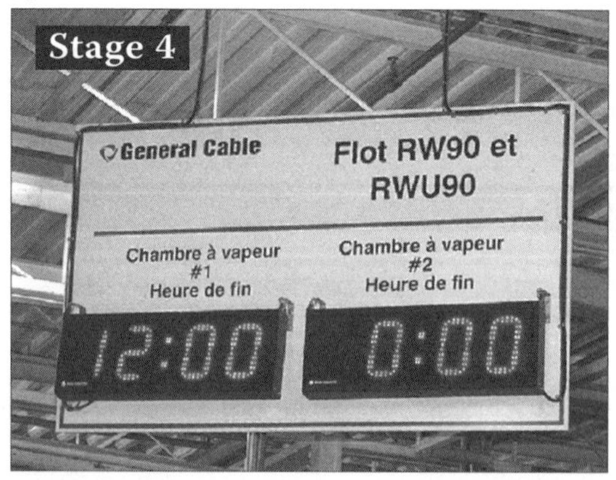

FIGURE 9.5
Stage 4: Alerts for upcoming gang changeover at constraint operation. (Courtesy of General Cable Corporation.)

FIGURE 9.6
Stage 4: Color-coded engineered WIP level. Note violation and how obvious it now is. (Courtesy of General Cable Corporation.)

FIGURE 9.7
Stage 4: Visual inventory signals, discrete color-coded locations. (Courtesy of General Cable Corporation.)

FIGURE 9.8
Stage 4: Red, yellow, green reorder points. (Courtesy of General Cable Corporation.)

FIGURE 9.9
Stage 4: Self-scheduling associates load Heijunka boards to Takt time. (Courtesy of General Cable Corporation.)

FIGURE 9.10
Stage 4: Associates well informed of value stream (i.e., mini-factory, progress, and results). (Courtesy of General Cable Corporation.)

The following section provides a summary of what is likely going on during each stage of visual management development.

STAGE SUMMARY

Stage 1

There is limited, if any, use of visual factory management techniques. Material replenishment by work orders happens in a traditional batch-and-queue mode. There is no evidence of kanbans, supermarkets, or flow lanes. Equipment, supplies, and materials have no demarcation and are randomly placed where space is available. Signage is poor. The plant is likely still arranged into process departments rather than by value streams.

The outside observer cannot easily discern the flow of products, bottleneck operations, or current status of customer orders and internal metrics.

Stage 2

This is still a traditional plant, but there will be evidence of experimentation in certain areas of the plant or with certain tools such as 5S. Replenishment is still driven by batch-and-queue work orders, but a kanban pilot might well be active or planned. Some signage exists, and there is some order given to where equipment, supplies, and materials are located. Aisles are generally clear of obstructions. Metrics are posted near work areas but may not be standardized or very visual—they are mostly numbers. Nonetheless, operators are beginning to understand what is being measured and why. They have a basic understanding of whether the business is winning or losing. There may be some evidence of cell pilots and the early stages of operator engagement.

Stage 3

There is significant use of visual management in the plant. Replenishment is mostly via pull signals using kanbans, supermarkets, or first-in, first-out (FIFO) lanes. Scheduling is evolving such that the constraint or bottleneck operations are the only ones being scheduled. There is creative and liberal use of color-coding, returnable containers, and flow lanes. Everything has a place and is mostly in its place. Operators are taking ownership of their work areas, with few exceptions. Locations are used to visually identify replenishment, to alert status, and to indicate issues. Andon signals identify the status of production and are used to alert assigned resources to problems on an as-needed basis. The constraint operation is visually obvious, and operators understand their roles in maximizing the bottleneck's throughput. Value stream maps are posted, and the line balance is clearly identified. There is evidence that kanban sizing is reduced periodically as process improvements are made and the buffer stock can be reduced. Metrics are now updated by the operators, are accurate, and are well understood.

Stage 4

Use of visual management is pervasive on the shop floor as well as in the offices. Replenishment of materials, both raw and work-in-process, is done visually. Kanbans, supermarkets, and FIFO lanes have been reduced to their theoretical minimums based on the process. The flow within cells has been optimized. Andon signals alert support resources as required to assist with problems that may occur and are beyond the operator's ability to solve. Color-coding, returnable containers, signage, and clear demarcation of locations support the flow that is documented in the value stream map. If something is not in its place, any operator who observes the condition will intervene for immediate correction. Metrics are visual, maintained by the operators, and used to manage the cells as well as to look for improvement opportunities. Operators behave like true business partners; that is, they take care of customers and solve problems.

> Visual management: It is 10 times harder to command the ear than to catch the eye.
>
> **Duncan Maxwell Anderson, American writer**

> There is no better zealot than a converted skeptic.
>
> **Larry E. Fast, founder and president, Pathways to Manufacturing Excellence**

10

Manufacturing Principle 9: Continuous Improvement

> Manufacturing Principle 9: Continuous improvement will result in relentless productivity improvement, year over year, forever. It is critical to the long-term job security of everyone to maintain a low-cost and competitive manufacturing cost structure.

We in North America, and especially in the United States, can sit around in conference rooms complaining how the playing field isn't level with, say, China and that we don't have a fair chance to compete. We can even hope that our government is working hard by various means to have foreign governments allow their currency to float or to stop government subsidies of their products. Where else in the world are the entitlement costs for environmental health and safety, for government regulations, for pension and health care costs as high as they are in the United States? What's fair is fair, right? Of course not. We must be good enough to compete regardless of all the obstacles. If we wait on the government to eliminate all of the inequities, they'll still be jawboning it with the competition on the day we close our last manufacturing plant. It's time to pull our collective heads out of the sand and find a way to compete. The very best plants continue to prove that it can be done by asking questions like these:

- How can we make the labor component more productive by removing unnecessary nonvalue-added labor from our processes?
- How can we control our processes to reduce waste to the theoretical minimum and increase overall equipment effectiveness (OEE) on constraints to at or near the theoretical maximum?

- How can we further leverage the supply chain cost advantage that we already have?

In every plant I've ever been in, whether it was award winning or doomed for closure, there was opportunity for improvement, albeit by different degrees. The poorest plant might have $100 bills all over the floor, whereas the best plant might have only $10 bills. In the 19 plants I've had a hand in closing over my 35-year career, the opportunities to get markedly better were both obvious and enormous. Unfortunately, the people who worked there simply did not have the will to improve after years of frustration. Usually, a plant closing is self-inflicted by a history of poor plant leadership, poor union management, and sloppy shop floor performance delivered by employees with poor morale. Why they have negative attitudes and "I am a victim" mentalities are subjects for another time. However, it is my view that over the years their management taught them to think and act that way. In my experience, this is typically at the root of why plants get closed.

Some of the more exasperating plant visits I've made over the years have been in plants where the plant manager and his team were so busy prancing around trying to convince me (and perhaps themselves) that they were "in alignment" with the strategy that they were walking right past $100 bills lying all over the floor. Clearly, they didn't even see them lying there in a pile let alone have any bias for action. It doesn't take sophisticated tools or problem-solving teams to bend over and pick the money up off the floor. Put in a quick fix and do the formal root cause analysis and problem solving after you've applied the tourniquet and stopped the bleeding. Let me cite a specific example.

In the wire and cable business there is a lot of commonality in the plastic compounds used to extrude the outer coatings (the jacket) of the cable. This is especially true for black compounds with common chemistry as well as other colors of compatible compounds that can also be turned into black. During a color change on an extruder it is common to have to "bleed" out the compound that remains in the barrel of the machine prior to feeding in the new color. These compound bleedings from the extruder crosshead can be collected, segregated by type, kept clean, and recycled right back into the machine with a regranulator. The best plants do this right at the machine where they have equipment that is dedicated to plastics with common chemistry. Others that have shorter runs or more of a variation in the kinds of compounds being used get most of the benefit by accumulating the material and later running it through a granulator off-line. Both can be cost-effective methods and allow the reuse of what would otherwise be scrap (or sold to a maker of floor mats for a few cents on the dollar).

So here's a plant that has a high cost structure but doesn't use reground jacketing compound. Why? "Because it doesn't run right," claims a machine operator. "Why not?" I ask. "Because every time I run it I have to slow the machine down and do extra screen pack changes, which causes extra downtime. To get the production out I have to use virgin material to avoid the loss of machine time." So I ask, "What do you find in the screen pack?" The operator replies, "Man, it can be nasty. There is dirt and sand, paper, tobacco, chewing gum, you name it." You get the idea. So what is going on?

Some operators are bleeding the compound that is left in the extruder barrel at the end of a run out onto a dirty floor. It is later collected and placed in a Gaylord box for processing. Others are bleeding it out into a container designed to collect this scrap. However, the containers have no lids. Also, the workforce at large apparently doesn't understand the value of the material and how it can be reused, so the container turns into a trash can. Coffee cups, candy wrappers, and gum all get tossed into the containers. Some of that escapes when the container is tilted to go into the granulator, and it is not seen again until it clogs up the screen pack. This is an example of what I mean when I say, "Bend over and pick up the money lying all over the floor."

This "process" happens day after day without intervention by an operator, supervisor, engineer, mechanic, manager—anyone. Who was it that said, "If you always do what you always did, you always get what you always got"? I'm also reminded of a paraphrase of Albert Einstein's definition of insanity: doing the same thing the same way and expecting a different outcome. These are the plants that eventually get closed. They can't seem to right themselves and at the end of the day deserve to be closed.

Thus, to drive continuous improvement, the plant manager must be relentless in his expectations for real productivity improvements every year, forever. Typically his engineering manager will be the point person for leading the cost-reduction efforts focusing on materials and process improvements. Meanwhile, the manufacturing manager will lead the way on the day-to-day controls while striving to meet the theoretical levels of scrap, material usage, and constraint throughput using engineering assistance for technical know-how and support.

When I tour a plant, I like to ask these kinds of questions:

- How is productivity measured? (Hint: If you can't find it on the income statement or balance sheet then it didn't happen. Here's where the finance manager becomes the key staff member to be sure we're telling each other the truth.)
- What is the plant manager's expectation for annual improvement? It's important that the leader understands and makes it clear to his team that there is zero productivity improvement until inflation (e.g., increased health care costs, general wages, material costs) has been offset. Productivity must mean year-over-year cost improvement net of inflation.
- How well aligned and intense is the workforce around driving cost reduction through process and materials improvement?
- How well are bottlenecks known and understood?
- How well is the theoretical universe of opportunity understood on material usage, scrap, constraint capacity, and OEE?
- What evidence is there that debottlenecking is a focus?
- Where is the impetus coming from for improvement?
- How well do engineers understand their role in this? Supervisors? Operators? Others?
- What evidence is there that cost reduction is a continuous process as opposed to just being an annual budget exercise?
- What percentage of savings is carryover from the prior year? How much has to come from new projects?
- What is the process for sustaining control of day-to-day performance?
- What evidence is there that Lean teams and Six Sigma Black Belt and Green Belt projects are an integral part of the process?
- Is there evidence of a robust process for internal benchmarking? External benchmarking?

Let me add a note on benchmarking—this is possibly the most sought after but underachieving tool in the box. Too many plants, in my opinion, get all flustered at the difficulty of finding good external benchmarks since that often involves their competition. This, of course, makes it impossible to do since you are no more welcome in their plants than they are in yours. Incidentally, some unethical suppliers sometimes like to offer certain "tips," which were never any comfort for me because I assumed they were sharing tips on my operations with my competitors as well. It's naïve to assume otherwise. Most of the equipment used to make wire and cable is common to the industry. If you have the money, you can have state-of-the-art equipment, but any competitive advantage is relatively short-lived. That's why the best plants find ways to differentiate standard, off-the-shelf machines by adding their own proprietary improvements.

Much of what sets excellent plants apart is how their associates create these proprietary advantages that one cannot see by taking a "main aisle" tour. My first plant to win the *Industry Week* magazine's Best Plant award in 2003 was an ignition set assembly plant in Altoona, Pennsylvania. The maintenance team in this plant is so creative and clever that associates would routinely build special jigs and fixtures that separated their plant's productivity from their best competitors. In fact, on a couple of occasions the business general manager called me and asked if we would allow a tour to a competitor that was also a current acquisition target. So we'd bring them into our Altoona plant. They'd walk down the main aisles of the plant and see the same brands and models of equipment they were using in their own plants. What they couldn't see were all of these small inventions responsible for helping the Altoona plant thump them in the marketplace. Now that's fun. As a group, the Altoona plant has the best and most creative maintenance crew of any I've had the pleasure to know.

I am continually amazed that plants generally don't even do a good job benchmarking in their own building. For example, what is the best practice being used in the plant to do a changeover on a 2.5" plastic extruder? If it is the third-shift operator on machine 4, then why isn't his process the standard work for all 2.5" extruders on all shifts? On occasion I've seen variation of more than 50% operator to operator. Please remember this: Before worrying about external benchmarking, benchmark your plant first, and make the best practices the standard work. Second, benchmark other plants in your company, and shamelessly steal their good ideas. Just don't forget to reciprocate as they will learn from a visit to your plant as well. Last, consider benchmarking external to your company. (Hint: You'll have to be at least Stage 3 before this will make any sense at all.)

Now let's look at the summaries of what is going on at each stage on the subject of productivity.

STAGE SUMMARY

Stage 1

Productivity improvement is an annual budget exercise. The plant regularly misses its numbers and requires assistance with project generation and execution. Costs typically increase year over year until competition takes large chunks of the plant's business until the plant closes. Very simply, plants cannot stay Stage 1 and survive.

Stage 2

Productivity improvement is formalized but is still mostly calendar-year driven. Engineers and managers are engaged. Project management tools are used but not widely understood. Modest cost improvements are made in some years and not in others. Internal best practice sharing and benchmarking are in their infant stages.

Stage 3

Productivity improvement plans are consistently met. Early signs of hourly associate involvement are evident. There are projects in the "parking lot" for initiation should any of the planned projects fall through or yield less benefit than originally thought. Gaps to plan are usually closed through this continuous, rolling 12-month process. Project management tools are widely used and understood. Internal best practices are routinely shared, and the plant manager sends his teams to other plants to shamelessly steal good ideas. The plant has matured in its culture such that productivity improvement is no longer viewed as just a budget exercise.

Stage 4

The productivity process is continuous and ongoing. Associates are constantly engaged in the pursuit of manufacturing cost improvements. Operators are involved and are often the source of project conception through the planning and execution processes. The facility is active in sharing best practices with internal facilities and has turned to external benchmarking to bring technology and productivity solutions from related industries into the company.

> We were fairly arrogant until we realized the Japanese were selling quality products for what it cost us to make them.
>
> **Paul Allaire, former president, Xerox**

> Many of life's failures are people who did not realize how close they were to success when they gave up.
>
> **Thomas Alva Edison, inventor and founder of General Electric (1847–1931)**

11

Manufacturing Principle 10: Communication

> Manufacturing Principle 10: A comprehensive, purposeful communications plan will be in place and executed in every plant on a rolling 12-month basis. Better informed associates take more interest in and make better decisions for the business.

This principle is based on the premise that management, however well intentioned, will do a less than adequate job of communicating with the right content and the right frequency unless there is standard work in place. The intent here is not to legislate the natural opportunities to communicate in everyday work life. (That said, most of those conversations could be much more purposeful than they normally are.) Rather, our purpose with Principle 10 is to define the minimum expectations for regular communications for all associates by various levels of the organization.

There are several reasons I think it is normal for leaders to often take shortcuts and do a poor job with communication. It is a disturbing reality that good communication very simply gets shuffled to the bottom of the priority list. Some will argue:

- I'm too busy right now to put any more on my plate.
- Meetings are expensive. The budget won't allow it right now.
- Meetings are a nonvalue-added waste of time. They are *muda*.
- I cannot afford to take the operators off the floor for a meeting. I need the production.
- We have all kinds of problems right now, and I don't have good answers for the questions I'm bound to get.

- I hate to speak in front of groups. (Polls over the years have cited public speaking as the biggest fear many people have.)

Of course, there are simple solutions for all of these concerns, but that isn't the point. The real issue here is that there is no understanding of the importance of regular communications with the leader's team. Therefore, there is no plan, no commitment, no understanding of the power of having a team in alignment and well informed about what needs to be done to deal with the issues of the time.

The need for this principle surfaced as I got into multiplant responsibilities and observed the broad differences in who communicated when, about what, and with whom. I found this odd because I had been trained from day 1 of my career to have regular staff meetings with my team, regular one-on-one meetings with my direct reports, and regular skip-level meetings so I could get unfiltered dialogue with hourly people in my department. I just assumed everyone did this kind of thing as routine. Man, was I wrong.

As I toured various plants I witnessed a host of communication approaches. Some supervisors had a standard "two rounds a day" routine when they'd walk the line with the operator and understand what kind of a day he was having. It was also an opportunity to see who won his daughter's soccer game the day before. Other supervisors would run to the problems of the day, which meant that their best operators might not see them for days at a time. Still others would tell their people they were always available if you want to come to the office. And how about those shop floor supervisors whose office is on the second-floor mezzanine? (Which supervisor would you like to work for?)

Staff-level managers also did their own thing with similar results. Some did a great job by having regular monthly staff meetings with their departments. These sessions were intended to keep everyone informed about how well they were accomplishing their mission for the plant. But more than that they kept all associates up to date with how the plant was doing overall and how the company was doing in general. It made associates feel part of something bigger than just their department. Other managers rarely left their office and couldn't manage the energy to get to the back of the office to see members of their group. They simply weren't in the main traffic pattern to and from the manager's office. Some plant managers spent scheduled time "dipsticking" on the shop floor by visiting certain work areas or shifts on a planned basis. Typically these same managers would host monthly "state-of-the-plant" meetings so that everyone understood what was going on with the business as well as specific wins and losses in the plant. The wins were celebrated, and the losses were put in context as to what areas of the plant would be responding to resolve the issues. There was also a weekly "stand-up" meeting where the staff would gather for 20–30 minutes and go over the game plan for the week highlighting any critical issues, customer orders, and plant visitors so that every member of the staff was on the same page.

Monthly staff meetings were religiously held principally as a "numbers review" but also included any changes in the game plan going forward. It was also a good forum to invite "hi-pot" management trainees to present or observe or to invite a division or corporate person who had insight relevant to that particular team. In contrast, some plant managers didn't leave the office much and, if they did, wandered around the plant without a purposeful agenda. Staff meetings, if they happened at all, were informal, and preparation for them was poor. They often turned into "bull sessions" that lacked any real agenda. Very little communication happened that gave a larger perspective to what was going on with the plant or the business on a more global scale. And, of course, there were variations all over the map between these two examples.

In the first chapter I mentioned the major acquisition by General Cable in 1999 when I oversaw 28 plants plus corporate staff. That significant development caused me to reassess my approach to communications since I would have less time to devote to it but a much greater need to make it happen. It was a priority to figure out how. I needed to revise my standard work and to add some new methods of delivery to accommodate the larger group.

First, I made monthly conference calls with my direct reports and all plant managers to update them on the state of the business overall as well as how the operations team was doing in delivering our part of the plan. We also looked forward and discussed any important current issues or potential changes to the plan. These calls were accomplished shortly after the monthly corporate leadership team meeting with the chief executive officer (CEO) so that I could convey the relevant news and insights to my team. The calls took about an hour including questions and answers (Q&A).

I also held monthly staff meetings with everyone on the corporate staff. Most were in the corporate auditorium listening to my direct reports and me report on the important business but at a slightly increased level of detail. Those who worked out of their homes or were on the road that day were expected to keep the time sacred on their calendars and to call in for the meetings. They were on the speakerphone so they could also participate in the Q&A. My assistant emailed copies of the slide presentation that would be used in the meeting so that they could follow along with the rest of us. The additional detail was required because many on the staff did not work with all of the metrics and terminology every day like some of us, and it was critical to educate and train the team to be sure we accomplished understanding. It was usually a very participative meeting with Q&A sprinkled throughout the meeting. I had a staff of about 50 people, many of whom were frequently in the field on various projects working with the plants. Those who didn't travel much still had regular contacts via phone, fax, or email. As a result, I always expected each member of staff to be up to speed with the priorities of the day so they could reflect our alignment throughout the organization at every opportunity.

One day a month was dedicated to the operations leadership team. I asked a couple of vice presidents (VPs) of manufacturing to travel to the corporate office for these meetings. While there was some cost associated with this, the guys were good at bundling other things in the office or in nearby plants to make the trip multipurpose. But the main point is that it's hard to build a team and communicate as effectively on the phone as it is in person. And this meeting was the most important of all from my perspective. This is when my team would give extensive reports of the key issues in their areas: which plants were doing well and which ones needed extra help and why; what part of our operating plan was on track or ahead (and why); and what part was at risk and required our intervention. What actions were required? Who would take the initiative and report back or get the resources required to put the plan back on track? Is there a need to reassess our plan due to significant, unforeseen changes? These were the leaders I depended on every day in the trenches to be well informed and aligned enough to instinctively do and say the right things. This is the group I trusted every day to put the ball in the goal.

When I oversaw 10 plants, I could visit each plant multiple times each year. With 28 plants, that wasn't realistic. I missed having the time to interact with lots of hourly associates during plant visits. But the organization was too big to try to communicate with several thousand people face to face. So we chose to do the next best thing: publish and distribute a DVD to each location once a quarter. These would be done shortly after quarterly earnings and other corporate news became public information. That way we could start with an overview of how the company was doing before going into operations performance and on down to how the plants were doing. We also developed simple visual scorecards to help communicate to hourly associates in a way that was easier to understand than all of the accounting and corporate lingo. Safety, quality, delivery, cost, and inventory were presented showing the plants grouped in columns of green, yellow, or red for each metric. The plant manager was, of course, asked to add his comments relative to his specific plant as well as to add his own content and do Q&A at the end of the session. The videos were usually about 20 minutes, and the total meetings typically lasted 45–60 minutes. Plants were expected to conduct the meetings within 4 weeks of receiving the materials. The corporate vice president of communications managed this process and followed up until each plant had completed the task every quarter.*

Plant visits were still important. I committed to personally visiting each location once per year. Even though I had manufacturing leaders and other members of staff in the plants on a regular basis, it was still important that I have a presence. Whenever possible I liked to have enough time to have dinner with the plant manager and his staff at a local restaurant. It was always a good way to help put people at ease and to take some of the anxiety out of "the big boss" visiting their factory. It was also a great way to get to know these important people for who they are when they aren't at work. It also helped to separate the personal from the business side of things when the next day's plant tour or staff meeting didn't go as well as it might have. The direct interaction was also extremely helpful when it came time for annual performance ratings, individual development plans, recommendations for promotions and transfers, or salary increases. I also loved to sit in on a plant manager's staff meeting to get their reports firsthand and to study the team interactions with the plant manager and among themselves. And, of course, a thorough plant tour was imperative to get a good calibration if this team's expectations and attention to detail were as they should be.

It was usually pretty easy to tell if there had been a major "clean-up" activity just prior to my arrival. Most of the time I'd just ask a machine operator, and I usually got the unvarnished truth. The funniest example was when I asked an operator that question in a plant in Kingman, Arizona. He smiled and pulled out a copy of a letter he had in his toolbox that had been posted on the bulletin board a few days before my visit. It was an exhortation to "get the plant cleaned up" before the boss from the

---

* Thanks so much to Lisa Lawson for being an awesome partner in this endeavor!

corporate office gets here on Thursday. It was signed by the plant manager. No joke.

Twice a year, summer and winter, we'd host a plant managers' conference for 2 to 3 days. Sometimes we would do this at or near the corporate office, but most often it was held at a pleasant location near one of the factories so that we could include a plant tour. This was the time that all plant managers could see one of their peer's operation for better or worse. I must confess, the operations leadership team and I often were purposeful with our choice of location. I cited the example earlier of having a conference at the Malvern, Arkansas, plant so everyone could see what the right expectations were for plant organization and housekeeping! The next meeting might be in a plant that had the highest score on that year's Manufacturing Excellence Audit (MEA) or a plant that was the model for their implementation of visual management. The networking time over a beer in the evenings was also a powerful benefit for plant managers to compare notes on problems, share ideas, and make dates to exchange best practices. It was also important that they knew members of their peer group well enough that they wouldn't hesitate to simply pick up the phone and call for help when they needed it and didn't yet want to involve their boss. It was a good forum for simply bouncing things off a knowledgeable and respected friend.

Every 6 months I'd have a skip-level lunch in a private lunch room with 5–7 corporate staff members at a time until I finished. My assistant would create the groups to be sure that we did not have anyone's supervisor in the same meeting and that we had representation from multiple groups. For example, I didn't want a lunch meeting with just purchasing people. The design was to have someone in attendance from, for example, purchasing, engineering, logistics, supply chain planning, and quality. The primary reason for this was to help break down the natural tendency for silos. I thought it was important that supply chain folks know engineers and that quality people know purchasing people. They needed to understand what each member of the staff did to contribute to the success of operations and ultimately the company at large. I sometimes heard, "Wow, I had no idea you did that. I always wondered about that," or "Now that I know that I'll get with you off-line because I think I can help you," or "Well how about that. You're exactly the person I need to talk to later to get some help," or "You probably didn't know this but I used to work in your area and I can help you with that." Similar to the plant manager conferences, these skip-level meetings accomplished multiple things.

The second purpose of these sessions was to field any general questions or concerns that they had regardless of the subject. I also made it clear that we weren't there to deal with individual problems they might be having with their supervisor. Any such discussions would have to be scheduled in private and that I wouldn't address them until after their supervisor had a fair opportunity to handle the concern. Often questions would come out of this smaller group that didn't surface during the larger monthly staff meeting in the auditorium. Associates frequently said they were too embarrassed to ask the question in front of such a large group or that they had been thinking about something I said, still didn't understand, and wanted more explanation about it. I also frequently had staff members come by and stick their heads into my office and say, "I'm going to be in Lawrenceburg (plant site) next week and wondered if there was anything you wanted me to check into for you." I considered every single member of my staff to be an extension of me in the field—another voice for excellence. It was therefore my responsibility to be sure that everyone in the organization was well informed and aligned. I must say, I just don't get it when people in leadership positions think these kinds of "meetings" are nonvalue added and not a priority.

I hope by now that you will agree that manufacturing excellence isn't likely to happen and has little chance of being sustained unless the normal randomness of communication is harnessed into a thoughtful plan and committed to standard work—thus, the communications plan requirement. And, as you might expect, the human resources (HR) manager typically owns this principle to coordinate across staff lines and to organize pulling together the plans in each plant with input from the staff and on behalf of the plant manager. (See Appendix E on the accompanying CD for a basic sample of a communications plan.)

I was especially fortunate during my 10 years at General Cable to have worked much of that time with human resources leader Peter Olmsted, who was progressive in his thinking and a highly supportive business partner in helping me lead the cultural change that was necessary in the manufacturing organization, top to bottom. His HR team leaders and plant HR managers were instrumental in helping to build the processes that would make our success in manufacturing sustainable through culture change.

I mentioned dipsticking earlier. I always liked to do a fair amount of that on plant visits just to get some unfiltered perspective of how hourly associates felt about their work, their leadership, their company. I always felt at a huge disadvantage having to speak through an interpreter in my Spanish- and French-speaking plants. Communicating through an interpreter just wasn't the same. But the point is that you can get a random sampling of what's going on in the plant from a communications standpoint by asking these kinds of questions:

- Would you mind telling me what your metrics are in this area? How well do you understand them? How well is your cell doing?
- How does your performance in this cell affect plant performance? Customers?
- How is your plant performing this month? This year? How do you know?
- What is your understanding of how the company is doing overall?
- Who are your plant's most important customers?
- Who are the main competitors for the products you make here?
- When was the last time your supervisor stopped by? What was the occasion?
- When's the last time you talked to your manager? What was the occasion?
- When's the last time you saw or talked to the plant manager? What was the occasion?
- What are the top three ways you are kept informed? (Typical answers are my supervisor, area bulletin board, union steward, small group meetings, large group meetings, personal visits, and home mailings to include the family and help to keep them informed.)

It's always interesting to see if you get similar answers to these questions in various parts of the plant. It's a good clue regarding who is and isn't effectively communicating. For example, if the associate knows a lot about how his work cell is doing performance-wise but knows nothing about how the plant is doing, then who isn't communicating? If the operator knows how well the plant is doing on overall delivery performance but isn't aware of a new job-posting policy that's been out for 2 weeks, then who is failing with communications? Nothing is ever this simple, but I think you get the picture.

Some of you may have heard this story. A man was walking along the street next to where construction workers were busy digging trenches out in the heat of the summer sun. The man approached one of the workers and asked, "What are you doing?" The worker's response was,

"I was told to grab a shovel and dig a trench. So I'm digging a damned trench." The man approached another worker and asked, "What are you doing?" The worker responded, "I'm digging a trench so we can pour the footers for a large building." The man thought to himself, "Well that's better." Then he approached another worker and asked the same question: "What are you doing?" The third worker replied, "I'm digging a trench to pour the footers that will one day support a beautiful cathedral." Now there's a worker who has the proper context for his work and therefore has much more pride than the other two workers, one of whom had no idea why he was digging and the other who came up short on the power of the vision and the importance of his work. It is human nature. People always feel better about their work and take more ownership when they are in the know.

Every worker in the business needs to have a context for his work, that is, how his work contributes to the grand vision of the company. If the workers are to understand they are part of the process of building a "beautiful cathedral," then their leadership must do a superb job of communicating that vision and why it's important—not on an episodic basis but on a continuous, seamless basis using the standard work of a robust communications plan.

The following section contains the audit key for Principle 10.

STAGE SUMMARY

Stage 1

There is no published communications plan and little, if any, formally scheduled communications. Operators are generally uninformed about the performance of their cell/department, the plant, and the overall business. Hourly associates rarely talk to management unless there is a problem and they are being targeted as "at fault."

Stage 2

A communications plan is under development, and parts of it may be on the way to implementation. It is not yet comprehensive, but it's a start. Hourly associates and staff salaried associates are becoming aware of what is happening in the plant. There is some sense that management is trying to improve communications, though there is still some skepticism as to whether it will be sustained.

Stage 3

The communications plan is much more inclusive and is evident throughout the plant. Bulletin boards are now standardized in terms of appearance and content and are updated regularly. Roles and responsibilities for the execution of the plan are clear. All associates are now aware of the performance of their work area, the plant, and the overall business. There is evidence of regular, open dialogue among associates at all levels. Associates are vocalizing how much better they feel about management's efforts to keep them better informed. Trust between management and the shop floor is growing. All associates are now routinely aware of visits by customers, suppliers, corporate officials, or other plant guests.

Stage 4

Operators now play a key role in the communications efforts within the facility. Communications flow freely up and down the organization. In meetings the communication is so seamless that you cannot tell who is hourly and who is salaried. Roles, responsibilities, and follow-up are clearly evident and well defined. Details regarding performance are readily available in the work cell and are being routinely updated, some of them by shift or by day. Hourly associates respect and trust their management team completely. Hourly associates play a major role in presentations and plant tours for outside visitors as a result of being so well informed.

I'll close this chapter remembering the best example of seamless communications I ever saw in a factory. It was in a small medium voltage cable plant in Moose Jaw, Saskatchewan, Canada, less than a 1-hour drive from the city of Regina, the provincial capital.

(Moose Jaw isn't the end of the world but you can see it from there. That said, it's a delightful little town with friendly, helpful people and a wonderful sports bar for watching hockey games. There is also a nice hotel, casino, and spa where the gangster Al Capone used to hang out when law enforcement was getting too hot on his trail in the U.S.)

The Moose Jaw work force was the best example of the culture represented by Operator-Led Process Control (OLPC) that I have ever seen (see Chapter 13 for more on OLPC). I walked into the plant late in the afternoon to spend some time with the plant manager and his team since I would have to leave to catch a flight shortly after lunch the next day. This plant had about 55–60 workers at the time, and the only person who worked in the plant that was not an hourly associate was the plant manager, Ray Funke. Even he had started out as an hourly employee when the plant had opened about 20 years before. Ray wasn't in his office when I arrived, so I took a walk out into the shop to find him and let him know I was there. I walked past some offices and saw Ray sitting at a long table. He was not at the head of the table but was just one of a half-dozen people who were talking about something. I slipped in the door as quietly as I could so as not to disrupt the discussion. I found out later that this group included machine operators, a quality technician, and a maintenance technician and a couple of the "staff" hourly "managers." Everyone was clad in jeans or casual pants and golf shirts. They all spoke with confidence and were all participating in a very balanced discussion about a production problem.

I don't even remember now what the subject was or the outcome of the discussion. What I'll never forget, however, is that after listening to the discourse for a few minutes I couldn't tell the difference between a machine operator and the "staff" member or the plant manager for that matter. The communication was absolutely seamless in terms of who was playing what position on the team. Nobody had any rank. It was a meeting of equals. The discussion was fact and data driven, respectful of everyone's input and focused on finding the best answer for the customer and General Cable. There was no finger-pointing. There was also a calm urgency to resolve the issue. That's the day the vision for OLPC became very, very clear to me. Shortly thereafter I added an exercise in a 1-day workshop that I do on the 12 principles of manufacturing excellence where teams go into breakout groups and try to visualize how the people are working and what they're doing on each principle: If you walk into the plant unannounced at midnight on Wednesday, what's going on in a Stage 4 plant?

> No organization can depend on genius; the supply is always scarce and unreliable. It is the test of an organization to make ordinary human beings perform better than they seem capable of, to bring out whatever strength

there is in its members, and to use each person's strength to help all the others perform. The purpose of an organization is to enable common people to do uncommon things.

Peter Drucker, management guru and author

The single biggest problem in communication is the illusion that it has taken place.

George Bernard Shaw, 1925 Nobel Prize for Literature (1856–1950)

12

Manufacturing Principle 11: Training

> Manufacturing Principle 11: A training plan will be in place and executed in every plant on a rolling 12-month basis. Fully competent associates work safely and deliver quality products, on time at a competitive cost.

Similar to the thinking behind Principle 10 on communications, Principle 11 is based on the premise that management often does not do a comprehensive job of training its associates. That could be during the onboard process for new hires; it could be when associates apply for and win a new job in the plant or when new information or new technology needs to be integrated into the training. It is not unusual to still see "shadowing" as the primary training strategy in plants. While this kind of training is better than no training at all, it puts the student in the position of being limited in his learning to whatever his trainer happens to know. Over time there is far too much variation in the training process, and we all know that variation is the enemy.

It's certainly not the appointed trainer's fault. My experience is that most people are flattered to be chosen as a trainer because it is a form of recognition of their tribal knowledge and because they are really good at what they do. But I don't consider this more than a small part of what it takes to have a truly comprehensive training plan, that is, one that not only has highly defined standard work content on the technical aspects of a particular position but also addresses safety and environmental requirements, company policies, basic Lean tools, and statistical process control (SPC) tools. And the list doesn't stop there. Training should address which systems impact on the new position; how to use them effectively; what the operator's role is before, during, and after a machine preventive maintenance (PM); which metrics are important and why and how to interpret them. It's also important to collect critical input from a variety of sources. For example, on the technical training, a team that included the machine operator, a quality engineer, process engineer, and a training expert from human resources might be the right group to get all pertinent input for the training plan.

Not only is training often inadequate, but also training budgets frequently are among the first to get the axe during difficult economic conditions. This is unfortunate and unenlightened in my view, but I also understand how tough some of the decisions can be when faced with making significant cuts. It's like advertising expense. So the boss rationalizes and says, "I know it is important, but I can't tie an outcome specific to a particular ad so I probably won't notice any impact if I don't do it for a year or so." In this environment what could be more important than having standard work for training "in the can" and available to whomever requires it?

One of my General Cable plants (Indianapolis) got creative and put a lot of its in-house training on DVDs featuring its best trainers teaching a variety of subjects. While it isn't as good as live training, recorded training is an inexpensive way of being able to carry on until the economic situation improves. Management simply has to take away the excuses that associates made costly mistakes because we couldn't afford to train them properly and they were ignorant of the requirements. Running training programs in a hit-or-miss fashion will derail any serious attempt to achieve manufacturing excellence and there will be no chance of ever sustaining excellence for any extended period in this kind of training environment. Here again, there must be standard work on all aspects of training and a cost-effective means of delivering it throughout the normal business cycles, that is, in both good times and bad.

As you might expect, the human resources manager takes the point on behalf of the plant manager and works directly with the plant manager and the rest of the staff to create the standard work for training. My enterprising and committed team at General Cable broke training down into bite-sized pieces and then over a period of a several years put a comprehensive manufacturing training program together. That progress continues to this day and will always be a work in progress as things change.

Training that was standard for everyone (e.g., safety, introduction to Lean, Lean tools training, the manufacturing excellence strategy, company policies) was created by corporate resources. Technical "how-to-run-the-machine" training and local policies were created and taught locally. Wherever possible the training was delivered in the plant training room to hold down the cost as well as to give it local ownership. Where the numbers weren't sufficient to support an entire class in the plant, regional training was hosted by a plant in that region of the country, and other plants within the company were invited to send attendees to fill out the class for a richer learning experience.

More sophisticated training such as for a Lean Six Sigma Black Belt would be held at the corporate learning center. This made it "special" for the participants. They got to see the headquarters, meet some of the corporate leaders and hear some of them involved with the training, develop relationships with their new peer group, and provide them with special recognition. Most of the training throughout the company was done by General Cable associates who had special expertise to share with their coworkers. Where expertise did not reside in-house, the training was outsourced.

As you can tell, there was a huge commitment to education and training at General Cable. I couldn't have been prouder of our corporate leadership team than during a huge downturn in the business in 2001 and 2002. America was in the backwash of the heinous events of September 11 and the colossal dot-com bust. We were closing several communications cable plants and treading terribly close to Chapter 11 bankruptcy. In spite of this headwind we continued to do Lean training. A leader of operations couldn't have asked for any more support than this. As you might imagine, it sent a powerful signal to the masses across the company that using Lean tools to enable manufacturing excellence was not going to be yet another flavor of the month. Lean thinking was here to stay.

While touring the plant, these are some areas to explore to get a sense of how the leadership thinks and acts about training:

- May I see the documentation of your training plan? (See Appendix F on the accompanying CD for outline of a sample training plan.)
- How do you address these requirements (e.g., safety, technical operating instructions, systems, behavioral and team interactions, Lean and Six Sigma tools, company and plant policies)?
- Who decides what skill sets need to be cross-trained, when, and for whom?
- Who decides the frequency and sequence of job rotations? How is that determined?
- How visual is the cross-training schedule? May we go see it?
- Will you please show me your operator certification process?

- How do you know what additional training is required for you to become a certified operator?
- Is "pay-for-knowledge" included in your training plan design? If so, please explain the thinking behind it and how it works.
- How flexible do you need to be to balance customer responsiveness with financial performance?
- How is the effectiveness of the training assessed? Which classes would you recommend to your coworkers? Why or why not?
- When was the last time the company canceled a scheduled training class?
- Typically who are the instructors for classes taught in the plant?
- Is there linkage between the individual development plans and the training plan for salaried associates?
- What percentage of the annual overhead budget is allocated for training?

Training must be a formal, disciplined set of standard work. What better way for plant leadership teams, human resources departments, and corporate leaders to actively demonstrate that they get it relative to Lean thinking and the need to build robust processes to deliver any important work—whether it's making the product or executing a training plan.

The following section provides the audit key for Principle 11.

STAGE SUMMARY

Stage 1

There is no published training plan and little, if any, formal training is happening. On-the-job training is pervasive as the accepted form of training. Multiple methods are used to do the same work.

Stage 2

A training plan has been developed and is in its early stages of standard work development. The outline is mostly in place, but there are huge gaps in available training content compared with what is ultimately needed.

Stage 3

The training plan is showing some signs of maturity. Cross-training is extensive in the plant to ensure coverage of bottleneck operations. Most operators can now articulate what skills they have and what additional training is required to be fully qualified. Job rotation has started in a pilot cell. Hourly associates are involved in the creation of training content and in its delivery.

Stage 4

Cross-training is extensive throughout the plant and operators are certified to perform at least two jobs where needed for flexibility. Operators understand how their function ties in with both upstream and downstream operations. Operators are involved in managing their own formal training requirements via a "training board" posted in the cell. Some operators are involved in the delivery of formal classroom training. They coach coworkers in both hard and soft skills necessary to work in a high-performance team culture (i.e., Operator-Led Process Control [OLPC]). Operators seamlessly communicate and move machine to machine as required to balance the line to meet current commitments to customers, shareholders, and coworkers.

> Man's mind, once stretched by a new idea, never regains its original dimensions.
>
> **Oliver Wendell Holmes, Supreme Court Justice (1902–1932)**

> You know what good leadership is? Tell 'em the rules of the game, train the dickens out of them and then get the hell out of the way!
>
> **Oren Harari, author, *Leapfrogging the Competition***

13

Manufacturing Principle 12: Operator-Led Process Control

> Manufacturing Principle 12: All associates will help to create and sustain a shop floor environment where the operator is in control of the process. This will be known as Operator-Led Process Control (OLPC).

The very first seed that was planted in my thinking for OLPC came from my dad. He was working second shift as a journeyman mold maker at the Overmyer Mold Company in Winchester, Indiana, in the 1950s. It was a regular occurrence for the family to have dinner together when Dad came home on his dinner break between 7 and 8 p.m. (His normal shift ran from 4:00 p.m. to 1:00 a.m.) While he hadn't been to college and was a very average high school student (he didn't like school), he was smart, a hard worker, and a creative thinker. He saw lots of things that could be done better, and he was often frustrated that his management ignored obvious improvement opportunities and simply coped day after day with dysfunction and inefficiency.

He used to say things like, "I can't believe what a dumb schedule we have in the Deckel machine department this week. Those yahoos up in the front office don't know their asses from a bass fiddle about how to schedule these machines. They make our jobs harder, not easier, and we're going to miss a 'hot' job this week because of it. If those guys would just stay in the front office the plant would run a lot better." I learned to understand that he was simply showing his frustration that he had no input on how to do his job more efficiently and effectively. Management in charge of important functions such as scheduling and time study was not seeking input from the operators. During my college years from 1965 to 1969, I worked summers in that factory, and I experienced first-hand what Dad had been complaining about. I remember thinking what a waste it was for Dad's bosses not to want his and other experienced operators' inputs. It was as if everyone not in management had been asked to check his brain at the time clock and just do what he was told.

The next major event that helped to form my thinking was in 1974. I was out of college, back from the Air Force, and 2 years into a 25-year hitch at Belden Wire & Cable. I worked in the production control department as a schedule supervisor. I had a "hot job" lined up in the plastic extrusion department to start "days only" that particular morning. Days only meant that this was a job that had been tried before unsuccessfully and would be started on the day shift under strict engineering control. The purpose, of course, was to have the technology expert, the department engineer, lead the parade and avoid another disaster for the customer and the bottom line. Because much, if not all, of the previous orders like this had ended up in the scrap barrel, the notion was to baby the process until sufficient quantities of shippable product could be produced.

That morning I arrived as usual around 6:30 a.m. and started my rounds in the plastic extrusion department to check on my hot job. This order was of a special construction. That meant that the order was for a nonstocked cable that had been designed for a specific application by a specific original equipment manufacturer (OEM) customer. It was a short order by wire and cable standards at only 5,000 feet. The order had been attempted twice and both times had been scrapped. The primary issue was that the order called for a jacket compound with which the plant had very little experience. You can imagine the expense (and the emotions) involved since the outermost layer of plastic (i.e., the "jacket" operation) was the last process. In other words, the cable was at nearly full value before it was turned into scrap.

This time the powers that be had decided to make 10,000 feet of cable, hoping that we would end up with at least 5,000 feet of quality cable to ship. The day before, I had verified that all 10,000 feet had arrived at the extrusion department, machine 15, and was flagged for "start days only" the next morning.

So it was about 7:00 a.m., and I had just spent a dime in the coffee machine for some of the hottest, foul-tasting caffeine sold in the county. As I walked up to the plastic extruder I was struck by what I saw. The engineers, to their credit, had actually shown up and prepared for a 6:30 a.m. startup to coincide with the shift change. This spoke volumes about the importance of the order because they normally started "days only" jobs when the engineering department reported for work at 7:30 a.m.

To continue with the story, two engineers and the department supervisor were hovering over the cross-head and control panel, and there was nothing of them visible from where I approached except elbows and asses. As I got closer, I could hear them giving each other orders of what to do next, and there was clearly some friction over the differing opinions. When I got into the work area, I looked to my left, and there stood the machine operator, a guy named Bill Darby. At that time Bill had been with the company for about 20 years or so, and for much of it he had run a plastic extruder. He was one of the best operators in the department, which is how he managed to get this difficult order scheduled on his machine after another operator had twice failed on another machine. It was the scheduling technique called, "If Bill can't run it then nobody can!" There was virtually no process data available that morning. It was a trial-and-error, dial-twisting exercise based on experience and tribal knowledge.

Bill had a very quiet and passive demeanor and was a real pleasure to know. He was one of those genuinely good people that you meet. He was a tall slender man in his 40s with dark eyes and still mostly black hair and beard. He looked a lot like Abe Lincoln. In fact, Bill enjoyed dressing up like Abe on Lincoln's birthday, top hat and all.

Bill was standing several feet to the left of his machine, leaning against a stand-up desk with his arms folded looking disgusted. When I walked up to Bill and said, "Good morning!" I could tell by his facial expression that it really wasn't. Mild-mannered Bill was irritated and upset. I asked him what was wrong, and he said, "These guys were already here setting up the job when I arrived about 6:20. I sure hope they know what they're doing." I said, "What do you mean?" Bill replied, "Well I put on my safety glasses and was getting ready to take over on the setup and discuss what we were going to try so we can get this order out today. The engineers basically ignored me and acted like I was in the way. I tried to give them some input, and it was clear to me they didn't want my input. So I walked away thinking to myself, 'Screw them. If they're so smart we'll just see how they do.' So I came over here to watch them and to stay out of their way—to see if they could make this order run by themselves."

I told Bill I was sorry that he'd been treated like that and that I'd see what I could do. Then I added, "Bill, can you make this thing run and get the order out of here today?" Bill smiled and said, "You get these turkeys out of my way, and I'll have this order out before noon." I did, and he did.

That's the day OLPC was born in my mind. That morning I vowed always to respect the hourly associates' know-how and to recognize that the machine operators are the only people in the plant who really add value in a manufacturing company. The rest of us are there to support them: all of us—plant managers, supervisors, forklift drivers, engineers, clerks, corporate staffers.

With that backdrop I'll try to paint a picture of what a Stage 4, world-class performing plant might look like in a culture of OLPC.

Principle 1: Safety

The operators are trained to do the job safely at all times. Operators proactively seek safer methods to do the job and are fully engaged in the development of job safety practices procedures and the training of new people who come on to their team. A formal review system is in place to provide a vehicle for identifying potential unsafe acts and conditions. Any near misses are fully investigated to eliminate root causes and prevent a repeat occurrence. Each operator is alert and keenly aware of anyone else who comes into the area to be certain that no unsafe conditions are created and that no unsafe acts occur. The operator has the responsibility to shut down a machine or stop an activity if the environment is unsafe. Safety is seamlessly and fully integrated into the culture. There are zero recordable incidents.

Principle 2: Housekeeping

Each operator keeps his work area in a 5S condition (sort, set in order, standardize, shine, and sustain). Everything has a place; everything is clean and in its place. This clean and organized environment is sustained without the need for management intervention. Housekeeping is inclusive of machine cleaning, painting, tool cleaning, and other minor maintenance. Operators routinely reassess the workplace for continuous improvement on sorting, setting in place, and standardizing the needed equipment and supplies. Common areas are divided up among the team members so there is responsible ownership for the entire area. Each operator routinely does his part without having to be told. Each person is careful executing these indirect duties while equipment is running. Each shift leaves the area exactly as they expect the shift to be when they arrive on their own shift. Any nonconformances are addressed immediately among associates during their hand-off discussions at the shift change.

Principle 3: Manufacturing Systems

Operators are fully trained on all of the systems required to do their jobs. This could include use of computer keypads, radio frequency scanning equipment, software, and completion of manual records. Transactions are always complete, accurate, and timely. Operators understand "the hooks" between their system responsibility and their upstream suppliers and downstream customers. Operators identify additional manufacturing systems tools and reports that enable them to perform better.

Principle 4: Preventive Maintenance

Operators routinely perform minor preventive maintenance (PM) such as machine cleaning, lubrication, and minor adjustments (e.g., tightening a loose guard, replacing a light bulb at the workstation). The operators also know their machines well enough to hear and see early warning signs of potential equipment failures, and they take the initiative to call their dedicated maintenance person to investigate before the equipment breaks down. PM schedules are then adjusted as appropriate. Operators are part of the maintenance team during scheduled PMs and during infrequent breakdowns to contribute their knowledge as well as to learn more about their equipment and its capabilities.

Principle 5: Process Capability

The operator is formally trained in statistical techniques and problem solving and routinely uses tools such as control charts, run charts, histograms, and capability studies, automated or manual, to control the process within upper and lower control limits. Statistical tools are well understood and routinely employed. Operators are fully knowledgeable on critical to quality (CTQ) and critical to process (CTP) characteristics of their products and processes. Operators seek process capability that is >1.33 Cpk as a minimum on all processes and 2.0, six sigma, on all CTPs. They routinely suggest process improvements that improve control. They are capable of diagnosing process issues and capable of solving many of their own problems based on the data at their fingertips.

Principle 6: Product Quality

Operators are responsible for product quality at the source and are well trained to follow the documented quality system. They routinely verify their own raw materials to ensure the correct input to the process according to the specification. They control their process. The output is delivered to the next operation or customer according to the specification. Operators understand the effects of substandard quality on downstream processes. Quality problems are most often solved at the source. If the process is out of control and cannot quickly be corrected, then the operator has the authority to shut down the process. In an OLPC environment the only scrap generated should be what is process inherent (i.e., the theoretical minimum).

Principle 7: Delivery Performance

Operators know who their customers are, how the products are used, and what products must be made in the current week and must control the sequencing to meet promised delivery dates. Operators are cross-trained to be multiskilled and flexible. They routinely rotate to other workstations to balance the flow, to keep constraints running, and to execute the schedules. Operators take the initiative to understand and correct the root causes of missed schedules in their control.

Principle 8: Visual Management

Visual management is pervasive and the primary means of managing the value stream. Operators routinely post and interpret their own metrics in the cell. Actual numbers are posted against specific work center or value stream objectives that tie to the approved budget and any stretch targets that have been established. Operators routinely manage kanbans, Andon lights, first-in, first-out (FIFO) lanes, and the like to control their scheduling environment and are alert to machine changeovers, equipment breakdowns, or quality issues. All operators stand ready to help a teammate as long as constrained work centers are not left unattended.

Principle 9: Productivity

Continuous cost improvement is part of every operator's responsibility every day. They take the initiative to present new ideas to the team and execute them if possible. If other functions need to participate, then the operator becomes part of a Lean team to pursue the improvement. The focus always is to eliminate waste moving toward the theoretical minimum scrap and material usage and to raise the theoretical output of the bottleneck. Operators are skilled with the use of basic Lean tools such as value stream mapping, fishbone diagrams, and Pareto analysis and use them as members of improvement teams. When the opportunities are beyond the capability of the operators, then engineering, maintenance, or other resources such as Lean Six Sigma Green Belts or Black Belts are dispatched as necessary to support the operator's initiative. These opportunities are resourced consistent with the size of the opportunity and the availability of shared or scarce resources.

Principle 10: Communications

When the plant has a comprehensive, purposeful communications plan in place on a rolling 12-month basis, then operators in the plant are conversational about the overall business conditions, about their own plant's performance relative to their objectives, and to their group of peer plants. Operators understand who their key customers are and whether there are any current issues with them. They are also aware of any customers or other distinguished visitors who are in the plant that day. They may be involved in making presentations in a conference room or at their workstation, or they may be serving as tour guides. Communications with members of the management staff occur seamlessly every day. Plant policies and procedures (e.g., safety, quality, technical, operational) are readily available to all associates at all times.

Principle 11: Training

When the plant has a comprehensive, purposeful training plan in place on a rolling 12-month basis, the operators in the plant are certified to perform at least two jobs other than their primary assignment. As product mix or volume shifts, they move to different workstations seamlessly to meet customer demand. When operators are absent, other team members make sure that the constrained work centers are covered first and that the flow is rebalanced as necessary. Operators enjoy the challenge of knowing multiple jobs and look forward to the variety that rotations provide. They self-manage their own training during the year to keep their skills current. Operators also receive appropriate training for their personal development consistent with their needs and interests. A formal, visual system is in place to track cross-training coverage to ensure that certifications are kept current.

Principle 12: Operator-Led Process Control

All managers, supervisors, engineers, and all other salaried and indirect hourly associates understand that their role is to support the direct labor operator. This is where the value add happens in a manufacturing business. They also understand that sustaining Stage 4 OLPC requires that all the standard work put in place on Principles 1–11 must continue to be done at a Stage 4 level everyday. All associates in the plant, regardless of their positions, communicate openly and work together to improve the business everyday working toward Stage 4 levels of excellence on all of the manufacturing principles. All associates are skilled in the use of basic Lean tools to find and eliminate waste with a sense of urgency. When all of these conditions are in evidence on the shop floor, then the operator is in control of the process and is delivering value to the customer and the shareholder. This is known as OLPC.

The following section provides the audit key for Principle 12 from the Manufacturing Excellence Audit.

STAGE SUMMARY

Stage 1

There is little operator involvement or interest in the facility or its performance. Overall plant performance is poor. Often a "one person, one machine" mentality is evident. Operators are expected to check their brains at the time clock and do what they are told in exchange for a paycheck.

Stage 2

OLPC is a topic in the training and communications plans, and there is some early evidence of operator involvement, particularly on safety and the start of 5S. Plant performance is on par with their traditional competitors.

Stage 3

Teams of operators are in place and functional, delivering results and meeting the business plan with consistency. Operators are increasingly involved across Principles 1–11 and are showing good ownership of their new roles. Salaried and hourly indirect associates understand their support roles to the direct labor operators. The use of Lean and Six Sigma tools is evident in the active Lean teams in the plant. All associates communicate openly and are working to solve problems and improve every day. There is a growing sense of urgency to get better. There is now a majority of hourly associates working in the new culture and OLPC has momentum across the entire plant.

Stage 4

Operators are an integral part of driving ongoing improvement in the facility. Their teams require very little management intervention. They are data driven and focused. Communications are so seamless that it is difficult to tell operators from managers. Management's role is principally coaching to assist them in accomplishing the goals. Associates effectively work in teams to control and improve their processes, deliver quality products on time, and improve the skills of their team members. They are customer focused and dedicated to keeping their commitments to customers, shareholders, and coworkers. There is a sense of urgency to continue to get better. There is disciplined, sustained excellence across all manufacturing excellence principles. Operators are in control of their processes drawing on technical and other resources as required for support.

> World-class performance begins with world-class trust.
>
> **Robert "Doc" Hall, author, professor, and founding member of the Association for Manufacturing Excellence (AME)**

> People in your organizations can cope with the truth, however unpleasant. In fact, they hunger for honesty and inclusion. If you're straight with them, they'll help you.
>
> **Oren Harari, author, *Leapfrogging the Competition***

Section 2

Leading Manufacturing Excellence

Readers should now understand the manufacturing excellence strategy as well as the thinking behind the four stages of manufacturing excellence and how they can be applied to each of the 12 principles. Now it's time to focus on whether the incumbent talent in the plant is capable of leading the cultural and performance revolution on the shop floor. That's what the second half of this book is all about.

Starting with the plant manager, I will share my observations and experiences. Then we'll go right around the leadership team table and review the positions that typically represent the staff of a manufacturing plant. (You may want to revisit the organization charts discussed in Chapter 1.) Does each functional leader understand his role in being the "thought leader" in the plant in his area of expertise? Is he up to speed technically on the state-of-the-art for his function? Does he possess the necessary leadership skills? Is he a team player who will seek functional excellence in a way that will be integrated with and supportive of the overall strategy? Does he have the intangibles necessary to stay focused, disciplined, and persistent? Is he totally in alignment with the direction established by the senior leadership?

A critical factor at this stage is that the senior manufacturing leader makes sure that he has people on board who think the right way. Does each member of the plant leadership team have the right technical skills as well as the people skills and the other intangibles to be successful executing this strategy? Of course this process starts by making certain that you have the right talent at the top of the plant structure. Our review begins with an expanded description of the plant manager's role.

> Perfection is not attainable, but if we chase perfection, we can catch excellence.
>
> **Vince Lombardi, legendary coach, Green Bay Packers**

14

The Plant Manager's Role

As you know from the preface at the beginning of this book, I highly value those who have helped to develop my thinking around manufacturing excellence. Absent their vision and input, I wouldn't have much to write about. Many bosses, teachers, and parents spend a lot of time barking out *what* and *when*. The closest many of them come to being helpful is to show us how to do things. But the most important and often missing part of these "superior to subordinate" kinds of relationships is *why* we do things—why they're important, why I'm being asked to do something that isn't experientially or intuitively obvious.

My experience is that a very small percentage of the managers I've observed over the last 40 years have been good mentors. It may well be that few of them had good mentors themselves and because of this they don't understand what mentoring means or realize the importance of it for their own staff's development.

Of the hundreds of supervisors and managers in plants and offices I have visited over the years, I have observed very few who really think the right way about their roles in the organization. They don't understand not only their own positions on the team but also how to best integrate with the other functions that are all important to the facility's success. That's why I'm going to devote this chapter on how to think about manufacturing excellence from the plant manager's chair. In subsequent chapters we'll look at it from the other chairs around the plant's staff table.

For the last several years of my career as an operations executive I presented a 3 × 5 card to each plant manager when he came onboard. In fact, I often used it in the interviewing process with both internal and external candidates to be sure they understood the difficult job for which they were interviewing. Further, it is critically important that they understand before they take the job how complete the accountability is for a plant manager.

> The corporate organization structure is a model that requires strong plant managers. Each must be a self-starter with his continuous improvement agenda; with an in-depth understanding of the business such that the right structure and the right people are deployed; with an energy level and an attention to detail that routinely delivers the forecasted outcomes; with the mental discipline, persistence, and confidence to effectively function in the matrix; and with the vision and leadership to revolutionize the shop floor using Lean and Six Sigma tools and to create an Operator-Led Process Control environment.

Over the years I have seen some outstanding staff and functional managers who have failed miserably as plant managers because they did not understand or handle the accountability that came with the job. Simply put, the plant manager is accountable for anything that happens or fails to happen on his watch, 24/7, 365 days a year. It's a no-excuses environment. It's worth noting here that it doesn't matter if you're working in a traditional organization or a matrix structure. Accountability at the top is without compromise.

Ideally, the plant manager is leading a world-class operation that is constantly anticipating unfavorable situations and is always preventing them from happening. But even Six Sigma companies fall out of bed three times in a million opportunities. Those three times the plant manager is accountable for objective root cause analysis and problem elimination. More typically, however, the plant is nowhere near world-class levels of performance and has frequent opportunities for the plant manager to be exposed. How he handles such situations in large part determines his future career at this level of management. Managers who make excuses or rationalize away the incident will fail. It is simply a matter of time.

The plant manager has a big job. I've worked with plant managers who had as few as 20 people and as many as 1,200. Regardless, the plant manager is the senior management person for the site. He is also the company's figurehead for the community in which the plant operates. As such, he's constantly in a fish bowl. He is the company in that community. His level of involvement, his communications style, his every move both inside and outside the plant reflects on his company for better or worse.

Obviously, the larger the city the more diluted the visibility. However, the workers inside the plant have the same expectations of their leadership as workers in a town of 2,000 people. They depend on their leader to set the example for integrity in all communications and in how the business is conducted. They want to be able to have complete confidence and trust in their leadership and to be proud of where they go to work each day.

All leaders who come into a company—regardless of their level in the organization, from the first-line supervisor to the chair of the board—start out with 100% credibility with their employees. It can only go down from there. That's why it is so critical that leaders protect their credibility to a fault. If you say you're going to do something, do it. If circumstances change and you don't or can't do what you said you would do, then at the very least communicate the change in direction along with why you made a different decision. People understand that things change, but expect it to be communicated in a timely manner. In my experience, not properly explaining why a change is being made is one of the most common failures of supervisors and managers. What is surprising is how many of them survive for entire careers by changing jobs without ever having had the trust of the many people who counted on them for leadership. And it's often really simple things that go unattended.

For example, an operator asks his supervisor if he knows what the schedule is for getting in a replacement part for his machine that will eliminate his current need to manually intervene during the packaging process. The supervisor says, "Sure, Joe. I'm heading to staff meeting later this morning, and I'll get an update on that project from the maintenance manager and let you know." The operator's shift ends at 3:00 p.m., and he hasn't yet heard back from his supervisor. As he heads home he gives his supervisor the benefit of the doubt, thinking he had gotten tied up on other issues and would likely get back with him first thing the next morning. Now it's 9:00 a.m. the next day and still nothing. Joe sees his supervisor walking down the aisle adjacent to his machine, but the supervisor doesn't make eye contact. Joe calls out to him but apparently isn't heard. After lunch the supervisor finally makes his rounds and arrives at Joe's machine. Joe has to ask what he found out yesterday on the delivery of the part. The supervisor responds, "Oh, man. I'm sorry Joe but I forgot to check up on that. I'll get back to you before the end of the shift." "OK," says Joe skeptically. On his way home from work later that afternoon, he was now doing a slow burn. Some of you may be thinking this is an extreme case. Based on my experience, unfortunately, it happens all too frequently and is at the root of how management credibility is eroded.

In all my years in leadership I never cared very much whether people agreed with my decisions, but it mattered very deeply to me that they understood why I took the position that I did. In my experience, although some groups of people didn't like some of my decisions, most at least understood why the decision was made. I know of very few who did not respect my position and my right to make the decision even though they may have disagreed with it. A leader can't ask for anymore than that over the long haul.

If as the facility manager we expect our employees to respect and perform standard work, to communicate openly with their coworkers, and to meet their commitments to the customer, business owners, and each other, then our employees need to be credible, too. If you agree on the importance of this for hourly and salaried staff, then how could we as leaders expect anything less from ourselves?

All of this simply means that the plant manager in a manufacturing facility is the senior leader at that location and is expected to think and act on behalf of the board of directors. And, of course, this means to represent what is best for shareholders, customers, and associates. It can be a real brain stretch for some who ultimately aren't able to think big enough to be a plant manager. But for those who can, the sky is the limit in terms of what they can accomplish with the help of their talented workforces.

A major role that the plant manager plays for his team is to interpret and explain the "big picture" to his team prior to directing action. Direction without understanding is mind-numbing and seldom gets the desired outcomes for the right reasons. (Remember from Chapter 11 the story of the three men digging a trench.)

How many of you who've worked in a factory have been called into the office one afternoon by the boss, who had just returned from a business team staff meeting? The boss sits everybody down and says, "OK, guys, I just got out of a meeting with the business team, and the vice president of manufacturing said that we have to cut expenses by 10% in our new year's budget proposal—something about sales going to be down. Don't argue with me because we don't have a choice. Let's just get it done. Meeting adjourned." That may be a bit harsh, but I've been in a meeting or two like that in my career where you had to stand up and start asking questions and hope to get something intelligible enough to translate it to your own team. Obviously, it's important that these sometimes necessary duties are carried out thoughtfully so as not to damage the business long-term. Some bosses, when confronted, will slow down and answer questions as best they can. Others lose patience because perhaps they don't understand either and may resent having to do it. How can this management team possibly communicate effectively with the rest of the plant?

The good plant manager will take copious notes in the staff meeting so that he can recap the highlights of why such expense reductions are important to the corporation. For example, he may highlight the year-to-year sales numbers and current year actual versus the next budget year's forecast. This will show the gap in real dollars and as a percentage, which can then be converted rather easily by the person on the finance team to how many dollars of lost operating income is likely to result from the loss of volume and possibly price.

Another example that could cause the need for significant cost takeout and layoffs is a sales drop driven by technology change. After the dot-com bubble burst in 2001, fiber optics was increasingly used to replace copper trunk cable to transmit telephone signals. It was much more efficient and lighter to handle. Those working in a copper telecommunications plant needed to understand the significance of that technology change so they could understand how critical the budget exercise was for their very survival. This kind of "sizing of the problem" communication from a "corporate perspective" doesn't make it any easier to do, but at least the team can understand what's going on as well as know how to field questions from their respective teams when they start work on the new budget.

The plant manager might then follow up by saying something like, "Once you've had some time to process this information (and maybe have some input you may have collected from your coworkers), please schedule some time on my calendar within the next week, and let's discuss your preliminary ideas. I am interested in seeing how you've approached this assignment and where you see opportunities in your area of responsibility. Please be sure to reach out to your finance counterpart to help provide the estimated impact of your ideas." That's real leadership and helps the team to pull together to accomplish a difficult but important assignment.

OTHER THINGS GOOD PLANT MANAGERS EXPECT

Anticipation

My Little League baseball coach, "Cricket" Lee, taught us always to anticipate that a ball would be overthrown and emphasized that we must always be in the proper position to back up our teammate. This, of course, was good coaching for us to prevent the opponent from advancing an extra base on an overthrow. I've found that to be a valuable lesson my whole life.

For example, if there is growth in sales that is beginning to push out lead times for customer orders, the production control manager (PCM) might anticipate that this might be the beginning of improved business conditions. After clarifying with sales and marketing, the PCM might anticipate that the plant manager should be made aware of this trend and should be presented with some alternatives to increase capacity and maintain competitive lead times. The PCM might also reach out and collaborate with the manufacturing manager, engineering manager, finance manager, and others for help in developing the most cost-effective way to increase capacity. In another situation, you may be walking through the plant and see a forklift operator stacking materials so high that you could anticipate that the load could become unstable and cause an accident. Your immediate intervention could prevent a coworker from injury.

A plant manager must surround himself with managers who will anticipate situations and cause a good outcome or prevent a bad one. If you're playing first base on the team and when the plant manager throws you the ball it frequently hits you between the eyes, then it won't be long until the team has a new first baseman. Anticipation is a trait that separates the mediocre from the best. That said, anticipation without action isn't worth much. Combine anticipation with initiative, and you'll get quick recognition as a potential talent worth developing for bigger things.

Initiative

Anticipation is recognizing the opportunity to improve the outcome. Initiative is the action you take without being told to make it happen. It is the bias for action. In the prior example about the PCM, he might have anticipated the need to do something based on the current trend of the burden being placed on the plant capacity. If he hadn't taken the next step (i.e., initiative to make something happen), the plant might well have missed numerous opportunities to grow the business and improve earnings because the extended lead times being quoted were no longer competitive and other manufacturers were enjoying the growth themselves with their shorter lead times. I have met zero outstanding leaders in my long career who don't have initiative in spades. They are the ones who are always making things happen. A plant manager should, of course, be setting this example with his own behavior and should expect no less from his direct reports.

Timely Communications

This is important both up and down the organization. Consider our example about extended lead times in the plant. How does the plant manager feel when he answers the phone to find a steamed vice president of sales on the line who wants to know why the plant isn't keeping up with growing sales and what the plant manager is doing about it? Very embarrassing. The plant manager and his whole team end up with egg on their faces and suddenly are in a reactive, damage control mode with one of the senior managers of the business.

Unfortunately, this kind of an episode sets off lots of wasted time as people try to distance themselves from the gaff instead of focusing on quickly solving the problem. In the meantime, the customer is not being taken care of and the business is being hurt. At this point it matters not how we got there but how quickly we can correct our course and take care of the customer. I used to tell my teams that "it doesn't matter who's responsible for where we find ourselves today. We're a team, and when one of us stinks, then we all smell the same."

Staff managers who know what to put on the boss's radar screen is a critical skill. Typically, the plant manager will discuss and will give examples of the kinds of things he wants to know immediately, such as an accident in the plant, a resignation of staff, an environmental breach, a significant problem with a customer's order or schedule, or a death in the family of an employee. But the staff manager must take the initiative (there's that word again) to initiate the contact. If the staff manager is bringing things to the plant manager that he really doesn't need to know right now, then the two of them can fine-tune their understanding. The rule of thumb is: When in doubt, communicate. In my experience the number one reason staff managers fail is because there are too many surprises coming out of their area of responsibility.

Feed Your Own Monkeys

I took a time management class early in my career—1977 as I recall—and one of the principle time wasters as suggested by the instructor was "feeding other people's monkeys." He used a *Harvard Business Review* article "Management Time: Who's Got the Monkey?" written by William Oncken, Jr. and Donald L. Wass in 1974, as the basis for teaching new managers a very important lesson. (My experience suggests that it's pretty wise counsel for more mature managers as well.) The lesson was this: You'll have staff people constantly knocking on your door and asking the question, "Have you got a minute?"

This should be your alert that your employee has a monkey on his back (i.e., a problem) and that his mission was to have that monkey sitting on his boss's back before he left the conversation. Of course, this was seldom a malicious act. More typically it was just the way the boss–subordinate relationship had evolved. In other words, the boss had trained his people to bring their problems to him for solution.

For example, a supervisor stops by his boss's office early one morning. "Good morning, Ron, do you have a minute?" The boss says, "Sure, Denny, come on in. What's on your mind?" Denny replies, "I've got a problem. I've got lots of orders coming up that require the same plastic compound, and I'm afraid we're going to run out before we complete this week's schedules. I think I'm going to be 2,000 to 3,000 pounds short." Ron replies, "OK, let me call Purchasing and see what they can tell me, and I'll get back to you in an hour or so."

Denny heads back to his workstation in the shop (probably oblivious to what he'd just done) but stops to visit for a while at the coffee machine in the department. What just happened here? Yep, the monkey that Denny had on his back when he walked into Ron's office had leaped off Denny's and onto Ron's back. (My guess is that Ron had quite a good inventory of bananas stashed in his office to feed all his monkeys.)

Now, let's rewind the tape and model a more appropriate response by Ron.

"Good morning, Ron, do you have a minute?" "Sure, Denny, how can I help you?" Well, Denny replies, "I've got a problem. I've got lots of orders coming up that require the same plastic compound, and I'm afraid we're going to run out before we complete this week's schedules. I think I'm going to be 2,000 to 3,000 pounds short." Ron's more enlightened response: "OK, Denny, have you talked to the clerk in raw material stores to see if there is any compound on the dock that hasn't been checked in yet? [Denny doesn't know.] Or is there another machine using the same material so we could call a priority? [Denny doesn't know.] Have you talked to anyone in Purchasing to see if we can get an expedited delivery in here in time to solve the problem this week [Denny doesn't know]?"

As you can see during this changing dialogue, Denny is being exposed for not having done as much as he could have done to help himself. This can be quite a good developmental situation depending on how well Ron handles it. Here's the rest of Ron's response:

> Denny, I know you need help but let's make this a developmental experience. Frankly, I expect you as a supervisor to solve all the problems in your area that you are capable of solving. Imagine how little time I'd have to do my own job if I took on every supervisor's problems in the plant every day. It would be impossible—just as impossible as if your hourly associates expected you to solve all of their problems for them every day. Let's take on the objective of getting you to a point where you can solve most of your own problems yourself. It's a real win–win for both of us and for the business. Now, Denny, as for today's issue with a possible shortage of plastic compound, do you see how capable you are of making the same phone calls I would have had to make to help you solve this problem? Of course, there is no reason you can't do this. So track down the alternatives I've suggested, and please give me a quick call once you've resolved the issue. If you still can't resolve it after you've explored all the options you and your team can think of, then by all means give me a call or stop by and I'll try to help.

Obviously, there may also be great value in this lesson for how Denny responds to his hourly associates every day. If he's too busy feeding their monkeys, then maybe he doesn't have time to do his own job. Assuming that hourly associates have been properly trained, my experience is that they normally know what needs to be done. Unfortunately, they often do not feel empowered to do it. They think they need permission. We must make it OK by showing our trust in them to do the right thing (see Chapter 13).

To sum up, when any of my subordinates tried to upwardly delegate their responsibilities to me, I was not happy. After all, isn't that what we mean with the feeding monkeys analogy: It is upward delegation pure and simple and is not acceptable behavior. Stopping this practice can have a huge impact on productivity in the business because it eliminates redundancy. Solve every issue at the lowest possible level in the structure of organization. The real bonus is that you get managers operating at the level where we need them to operate to improve the business and make more money. Isn't that the goal (see *The Goal* by Eliyahu M. Goldratt and Jeff Cox, North River Press, Inc.)? You also take full advantage of using the brainpower of the hourly associates and help them to contribute more and to feel better about their jobs. It truly is a win–win–win.

Develop Your Own Organization

The previous example demonstrates the importance of each manager developing the talent in his own organization. Most companies have a formal process in place to measure performance of the salaried staff against annual objectives. This often is tied directly to the compensation plan. It also normally includes a personal development instrument designed to include associates' work experience and formal education and professional associations as well as to identify strengths, weaknesses, and the incumbent's career interests and ambition. The most important part of this process is the development plan. That's when the associate and his supervisor have an open and honest conversation about the content of the forms themselves and what specific personal growth objectives they will target for improvement in the new year. This could be a college course, a seminar, or some in-house training to beef up a particular skill set. Some examples are time management, how to discipline associates, a refresher on company policies, and statistical process control.

This interview may also surface the need for improved soft skills such as communication with associates. For example, it can be very helpful to take a battery of Myers-Briggs and FIRO-B tests to improve one's understanding of his DNA so he can learn to change his behaviors and improve his interaction with others.*

The key point is this: The manager and subordinate enter into a contract to work together over the next year executing the agreed-on development plan. The associate takes the initiative, maybe with the help of the human resources team, to find the right content and format to address the area of development need. Then the supervisor follows up formally (e.g., once a quarter) to make sure that the development plan

* While it is beyond the scope of this book to discuss specific tools in detail, these are important team development tools worth sharing. I can personally attest to their value, having attended a wonderfully eye-opening and helpful experience called Leadership at the Peak, conducted by the Center for Creative Leadership. The use of these tools is especially effective when you do it together as a team with, for example, the plant manager and the entire staff. This helps each member of the team learn to use the strengths of each team member while also understanding their "flat spots" and helping them to compensate for an area of weakness. It also helps to take the sting out of criticism when a teammate can say, "Hey, your FIRO-B is showing. Let me help you out here." It can be nonthreatening and even fun as it creates a whole new dynamic within the team.

is underway as agreed. On a more informal basis, the boss can also give the associate feedback on a regular basis by saying, "Hey, Tim, I've really noticed how much you're trying to participate more during our weekly staff meetings. I hope you've also noticed that your teammates respect your input and listen carefully when you speak. This is the kind of behavior we're looking for in your development plan. Keep up the good work."

If every manager and supervisor in the plant would give this kind of time, attention, and care to each of their direct reports, can you even imagine how much stronger your organization could be? I had the pleasure of leading plant managers for 25 years, and my expectation was that each one should develop his organization such that when anyone leaves the plant the replacement is already in the building and is either already qualified or is close enough to avoid an outside hire. Ideally, the only new hires into the salaried ranks should be for entry-level positions where there is not an hourly associate in the plant who is capable of stepping up. Note that the leader always reserves the right to bring in fresh thinking into a key position if his judgment is that there's a need for special skills or to supplement mid or upper management needs.

I admit this is a formidable objective. I've seen a number of plants that had good track records on organizational development, but I've seen only one plant that actually accomplished this objective over a period of 15 years or so. That was the Indianapolis Compound Plant of General Cable Corporation, which was an *Industry Week* Best Plants winner in 2007. The man who was committed to that goal and made it happen was a fellow Hoosier, Dan Jessop.

Properly Administer Company Policies

Before starting this section let's be clear on one thing: It is the company's responsibility to provide adequate training regarding company policies and procedures and to regularly update the entire salaried organization when anything changes. Otherwise, it is not fair to expect that managers and supervisors are capable of properly upholding the policies of the company. That said, anyone playing a supervisory role represents the company to the hourly associates they supervise. This is especially true when it comes to company policies.

When associates have questions, they expect their supervisor to know the answer or to know exactly where in the organization they can go to get the answer. For example, no supervisor would be expected to be an expert on the company's health benefits, but he sure ought to know the name and the extension number of the person who is the expert. On the other hand, the supervisor should thoroughly understand the policy detailing how overtime work is distributed for hourly workers.

It's also important that the supervisor understands the intent behind a policy so that he has a thoughtful response to the associate who asks, "Why is overtime awarded based on having the most seniority in the classification? I've been here longer than some of the machine operators who are working overtime this weekend. I should get the opportunity to work the overtime, too. Doesn't department or plant seniority mean anything in this company?"

If the supervisor responds with something like, "Joe, that's just the way it is, and it's been that way since way before I got here. You know that. You've been here as long as I have." Well, that's an answer, OK. But it's more likely to anger the associate and is certainly a good example of poor communication. In this case the supervisor has abdicated his management responsibility and hasn't properly explained the policy. Let's try a more enlightened response from the supervisor, Dan, in the following exchange:

Dan: Joe, I understand what you're asking and I know you've been here long enough to know that this isn't a new policy. It's been in place longer than we've both been here. But let me give you the thinking behind this, and then we'll see what we can do about your particular situation. We're scheduled to work overtime because we have important customer orders this week that will not ship on time unless we work on Saturday. The company will be paying the time-and-a-half rate, that is, extra labor cost, to make the deliveries. Let's assume that we allow someone who has more department or plant seniority to work the overtime. What are the implications to that? Now we're not only paying a premium for the labor, but we have an untrained person running the machine which likely will result in less production and higher costs such as downtime and scrap. It just doesn't make good sense to run the business that way. We'll miss important customer deliveries and hurt the bottom line in the process. Is this making sense?

Joe: Sure is, but I could sure use the extra money right now. I've got a daughter in high school now who wants to go to college, and I need to start making some more money.

Dan: OK, why don't we talk about how you can get a job in one of the departments that's forecasted to get a lot of overtime, for the foreseeable future. Another one of our company policies is to promote from within whenever possible. Let's see how it might help you get what you want. We post every natural opening in the plant for all hourly classifications. You can use your plant seniority to bid on any of those jobs. With the number of years you have with the company it's pretty likely that you'd have a good shot at some of those jobs as they are posted. Let me remind you that job postings for the whole plant are posted on our department bulletin board over by the time clock. There are also bid forms there you can fill out and drop them in the box. A representative from human resources, normally Sally Jones, will stop once each morning to collect the bids from the box, take down postings that have been filled, and post any new openings that the plant has. I'd suggest you make that bulletin board a regular part of your daily routine so you don't miss any opportunities to get the new job you want in the plant. What questions do you have, Joe, that I might help you with?

Joe: I'm good. It's been years since I've changed jobs here, and I'd forgotten a lot about how the process works. I'll start working the bulletin board every day when I'm on break and on the way in and out of the plant."

Dan: Sounds good. With all of your experience I'd sure like to keep you in the plant somewhere even though I'll miss having you in my department. Good luck. I hope this works out for you. Please keep me informed where you are in the process when I make my rounds.

Joe: I will, and thanks for helping me get my head straight. I'll keep you posted on what I bid on and how the process goes. If I do land something in another area I want to give you as much heads up as I can so nothing falls apart for you during my transition.

Some readers may be thinking, "That's sounds pretty simple if you're in a nonunion plant, but I can tell you it makes little sense in a union plant. How jobs are awarded is in the labor contract. It's cut and dried." Nonetheless, even if the rules are spelled out in the contract, my experience is that associates who haven't used the process in years simply haven't kept up with it. Further, regardless of whether the plant is union or nonunion, the hourly employees work for the company, not the union. I've never wanted a third party to become a substitute for open and honest communication between supervisor and worker. There could well be additional steps needed in a union environment; for example, a union representative may have to sit in on the conversation Joe and Dan just had. But that's OK. It's a way of building trust in the plant. It demonstrates concern and commitment of the supervisor (i.e., company management) to help guide an employee to an improvement opportunity. Also, the next time he has something on his mind on any subject Joe will be more likely to approach his supervisor.

Plants that can create this kind of real connection with their workforce have a chance to be something really special in terms of performance. They also have a much better chance at creating long-term job security.

Make the Numbers

This brings us full circle with regard to the plant manager and his staff being held accountable in a no-excuses climate. His staff must understand that their collective job is to forecast the expected outcome and then to do whatever is necessary to make it happen. Some plants get in the habit of just reforecasting the outcome when they hit a bump in the road. Over time this creates a culture that it is acceptable to not make the numbers. If that kind of thinking by the leadership doesn't get nipped in the bud it is the beginning of the end for that management team, if not the plant itself. One of the former chief executive officers for whom I worked, Stephen Rabinowitz, said it best: "Our job is not to just report the numbers we get. Our job is to make the numbers that we seek."

My entire 35-year career has been with public companies, so that is the context of my comments on this topic. While I'm sure private companies have similar expectations, they are not under the intense scrutiny of public companies. Shareholders in public companies have relentless expectations to earn a fair return on their investment in your company. Failure to deliver on promises ultimately causes them to sell their stock and seek investments elsewhere that are more dependable and lucrative. Failure to perform over a period of time always causes the share price to drop through the floor and destroys the best currency that a public company should have for growing the business.

Something else is often misunderstood by many, especially those who are in the organization a few layers below the executive leadership team.

When budget time rolls around each year there is frequently a misconception that the leadership team has pulled some ridiculous numbers out of a hat to use as impossible targets for the new year. (Come on now, folks—I know some plant managers and others have thought this and worse.) Nothing could be farther from the truth in most cases. I've seen situations where the new targets have been as arbitrary as "the Wall Street analysts expect us to grow sales and earnings by 15% per year or else they'll trash our stock." In those unfortunate cases, the targets ignore lots of economic factors that in some years make those kinds of numbers unachievable. Nothing is more frustrating for a team than to have objectives set that way. At the end of the day the stock will get trashed anyway when the numbers that have been publicly committed are not delivered.

I certainly don't intend to imply that there is a lot of sympathy at the board of directors level or among leadership team members. Both groups are charged with taking on challenging goals to help the company grow sales and earnings. (That's what everyone in management gets paid to do, isn't it?) With the help of the leadership team, the board will look at the macroeconomic indicators and perhaps communicate a range of growth expectations by regions in the world. The leadership team may define it slightly finer by breaking it out by product group so that the mix of sales and earnings is clear. This helps to have an early look with the assumptions as follows: If we grow only at the averages being forecasted, what will be the forecasted outcome on sales and earnings by region and for the full corporation? What are the opportunities to do better than that by country and product line? By how much? What are the risks that the outcomes could be worse than that? By how much? There would likely then be some handicapping done to arrive at a set of targets. These targets are then communicated through the leadership team so they can begin to determine what impact their organizations can have to deliver a reasonable outcome. *Reasonable* includes some degree of risk. "Make-the-numbers" companies don't sign up for "slam-dunk" budgets. They want the organization to have a challenging yet reachable goal that optimizes the outcome for investors, customers, and employees.

Rather than make a short story longer, suffice it to say that a lot of topside thought goes into the creation of macro objectives. As these get translated into specific product demand forecasts for each plant, then the plant manager and his team get heavily involved in the development of the new budget (i.e., business plan) for the new fiscal year.

Now let's assume that the plant manager and his key staff members have all been involved in vetting the proposed sales forecast with members of the business team and product management. Further let's assume at this point that we in the plant have agreed that there is plant capacity and material supply available (or can be made available with the correct timing) to make the volume numbers with the product mix being forecasted. Obviously, if this isn't yet the case more negotiation is required to shape the volume and mix into a plan that is executable. Signing up for a plan that is not possible to execute simply disappoints customers and shareholders and is often the shortest distance between being employed or not for a plant manager. Have the tough discussions now, not after you've already disappointed key constituents.

Once there is a path to deliver the volume and mix being forecasted, the plant manager now must understand the business team profitability objectives and determine how his operation can contribute to the bottom-line expectations. For example, what is the pricing environment in the marketplace? Will cost reductions be required just to offset price erosion in the marketplace, or is it likely that price increases will lead improvements in profitability? From the plant manager's perspective, year-over-year cost reduction is required regardless of the market environment. That said, understanding the market environment certainly provides the clues to the necessary pace of improvement, for example, whether capital spending can be part of the solution.

Armed with strong input from engineering, finance, materials management, and manufacturing management, the plant manager will likely lead a series of what-if scenarios. This is important to ultimately determine a plan that has some stretch in the objectives but is deliverable with good cross-functional coordination, project planning, metrics, and follow-up. If possible, all of the individual cell or department measurables will add up to the plant's deliverables for the new year's business plan.

Now let's suppose that one of the important objectives is to deliver a 2% improvement in overall equipment effectiveness (OEE* = machine utilization × efficiency × first-pass yield) for the plant. How many hours of useable capacity, by work center, will such a change provide? How many dollars of improvement will result and ultimately show up as additional margin contribution for the business? What do we need to know to track

* OEE is for machine-paced processes. If the process is paced by labor e.g., manual assembly, then calculate OLE substituting labor utilization and efficiency for a similarly powerful metric.

the progress hour by hour to ensure a "hard-wired" outcome versus "the plan"?

First, we need to know what the equipment constraints are to deliver the promised volume and mix. For a 2% plant improvement we may find that a 10% improvement is necessary on certain constrained work centers. That being the case, the manufacturing and engineering managers must have a path to deliver the 10% capacity increase on the targeted constraints. With help from their staffs and with input from hourly associates, a project list is developed that on a calendarized basis will deliver the necessary results. (Note to the human resources manager about training: Formal project management skills should be present for those charged with delivering the project results.) Most importantly, everyone associated with these projects must understand that improving capacity on nonconstrained work centers yields no improvement at all to the business unless it is sufficient enough to reduce headcount. Otherwise it is very low-priority work, which should not be resourced until the constraints have been moved by superb execution of the high-priority projects.

Now on to the metrics. Once we have a good handle on which work centers and which projects are critical to delivering the plan, the metrics must be in place so that the shop floor personnel know hour by hour how well they are doing compared with the planned expectations. I have seen plant managers make two big mistakes at this juncture. First, projects are launched without being sufficiently vetted and understood. The unsatisfactory result is found later (often too late to make up the deficit) when the project manager says, "Yeah, boss, I've completed that project," but the finance manager can't find the results anywhere in the numbers and the plan is being missed. The plant manager and staff should invest whatever time it takes up front in the budgeting process to make sure that the project manager has a clear and measureable path to the promised outcomes. Going forward into the year, as circumstances change, the projects must be revisited to make sure that there is still a path to making the numbers.

The second biggest mistake typically made is that the metrics are set at too high a level. For example, the OEE improvement number might be reported weekly at the focused factory level, the cell level, or the work center level. My view is that while knowing these things may be relevant on some level, the one nonnegotiable metric is to know it at the machine level by the hour. We can't afford to find out we have a problem the next day or at the end of the week. We need to know as close to real time as possible so that we can intervene as required to change the outcome to the plan commitments.

There are two very simple and particularly effective ways of dealing with this. First, put a clipboard next to the machine along with a sheet of paper that shows the 24 hours in a day (or less if the machine isn't scheduled to operate around the clock) and three markers—one green, one yellow, and one red. Then discuss with the operators the purpose of this measurement tool and that you need their help in making sure we know how we're doing delivering our commitments on this critical machine. The operator who has been properly trained and has the proper shop floor packet for the job he's running will know in real time whether the machine is running at the rated speed in the standard, whether setups and changeovers are being done in the prescribed time, whether any raw material or maintenance issues are affecting efficiency or utilization. As long as things are running to plan or better, the operator simply colors the space for that hour in green. When an issue puts machine performance at risk, the operator colors that area yellow. Within the next hour the operator should have corrected the issue and be back to green. If that is not the case, the operator colors that hour in red and enlists the help of his supervisor, maintenance, whoever is required to get the machine back in green operating condition. This also makes a great visual management tool as the supervisor or the assigned engineer or maintenance person makes his rounds. You can't have a good day unless you get it done an hour at a time.

A second easy way to communicate performance on constraints is simply to provide Andon lights for the operator to use even before he's posted his job status into the formal system. This is as simple as installing three lights—the same green, yellow, and red as before—with a switch close to the operator. The lights should be high off the floor and visible from across the area (e.g., to a supervisor's office, maintenance area). He then simply turns on the appropriate color light to reflect his current operating condition. One could also install an audio alarm that goes off or put a strobe on the red light so that it gets instant attention when a constraint is completely down or in serious trouble due to excessive quality defects or mechanical breakdown. The constrained work center that is down should look like a beehive with the appropriate resources necessary to get it back into operation quickly. Any work on nonconstraints should cease immediately if those resources are required to resolve the issue with the constraint.

The final point is to be adaptable. No plan is ever constant and correct. On the contrary, nothing ever goes exactly to plan. Forecasts of volume and mix aren't perfect, and neither is performance in the plant. Consequently, it is as important to recognize when you need to run up the flag to change the plan as it is to have a plan to begin with. Whether the revision represents good news or bad news, it's important to initiate discussion about the need for a revised plan as soon as possible. Keep in mind that until and unless the leadership team and the board of directors have reforecasted the company's performance to shareholders, everyone is still committed to making the original earnings that have been approved and externally communicated. Given the new circumstances we simply have to find a different path to the same outcome. This can often lead to frustrated staff members and claims of how unreasonable the senior leadership is. This is where the plant manager has to be at his best in keeping everyone in the know about the changing market conditions and the collective need to recast the path to the plan.

I hope by now it is crystal clear how big a job the plant manager has. He is the on-site representative of the leadership team and the board of directors. If the plant manager isn't thinking big, isn't involving and educating his staff to think at a high level, and isn't making his expectations clear, then the staff will not properly respond when the original plan is put together or when it goes south. Plant managers must be mentally tough and make the difficult business decisions that they get paid to make. Those who are preoccupied with being everybody's buddy will fail in the long-term. There simply has to be a certain level of "separation" so that personal relationships don't get in the way of making the right decisions for the business.

I'm reminded of a plant manager who had such a strong personal relationship with a member of his staff that he could not or would not recognize how incompetent his friend was in his current role. And his friend was responsible for much of the chaos that was present on the shop floor with no end in sight. After numerous counseling sessions with the plant manager on this subject over a period of months I had to take a position with him: "If you will not deal with this issue, then you leave me no choice but to remove you first so I can deal with the problem." As difficult as these kinds of things are, the board of directors is not paying any member of management to allow personal feelings to get in the way of doing the right thing for the business.

Plant managers who get defensive, are argumentative, or are slow on the trigger to recognize the need, to communicate it, and to respond with replanning are plant managers on their way to whatever's next in their careers. The best plant managers are those who stay objective and focused on making their commitments, set the right example for staff, and are appreciated, low maintenance people in their boss's world. These are also critical reasons the plant manager must surround himself with the best talent he can find for the important staff leadership that will ultimately define the plant's success or failure. The next challenge is to make sure that each staff manager understands his important role as a member of the team and plays his position effectively. We'll now explore those positions one by one.

> Delegation without follow through is abdication.
>
> **Larry E. Fast, former senior vice president, operations, General Cable Corporation**

> Sgt. Preston's Law: If you aren't the lead dog, the scenery never changes.
>
> **Lewis Grizzard, author, *Life Is Like a Dogsled Team***

15

The Manufacturing Manager's Role

In a traditional factory, other titles for this role are production manager and plant superintendent; in a cellularized or focused factory setting, it may be value stream manager or unit manager. The manufacturing manager is the person who is responsible for all of the direct labor associates in the plant, their supervisors, and indirect support people. He basically owns responsibility for the shop floor. Typically the person in this role, by position, is considered the number two person in the plant management structure. There are, of course, lots of exceptions based on the strength, leadership, and upward mobility of an up-and-coming staff member—materials or engineering management are the most common.

So why is this role normally perceived as the number two spot? In a manufacturing company you will typically find that the manufacturing organization manages about 75% of the people and spends 75% to 80% of the money. Further, since manufacturing is where most of the value is added in the company, indirect labor and salaried staff are in support and staff roles to help ensure that direct labor functions as effectively as possible. Of course there are many examples where the manufacturing manager has settled into a career position due to his lack of capability or interest in assuming the full accountability of the plant. This is OK as long as the incumbent is the continuous improvement zealot on the shop floor and is well read and practiced with key Lean and Six Sigma skills so he can effectively lead "the revolution." Without this mind-set and these skills, he blocks progress and organizational development and must be managed out of this role.

Over the years, when I was looking for future plant managers, the manufacturing manager's chair is where I always looked first. No other position who reports to the plant manager typically has the same combination of product, equipment, and scheduling knowledge—a huge edge in being able to lead, direct, and educate a group of supervisors to improve the business. Outstanding manufacturing managers also have to demonstrate that they can take the heat and be accountable for a huge part of the business. But the biggest advantage is the fact that this manager, if he's worth his salt, knows most if not all of his hourly folks on a first-name basis and has earned their respect over a period of time. When the manufacturing manager gets promoted to plant manager, the shop floor reaction should be predictably positive. The kind of thing one should overhear from hourly folks in the break room is, "I'm glad to see that Mike got promoted to plant manager. He's a smart guy, and he really knows his stuff out here in the factory. I also like the way he always listens to our ideas and involves us in key issues. I've got a lot of respect for him, and I think he'll be a good one."

The alternative is the staff manager who hasn't yet had the opportunity to establish the same base of support in the hourly ranks and is an unknown entity to the masses. The hourly response might go something more like, "Who the hell is Ernie? I hear he's the materials manager, but I seldom see him out in the shop; it seems like every week I'm running out of one material or another at my machine. Besides, how can he be a good plant manager when he hasn't spent time in the trenches with us?"

This reaction is also predictable even though Ernie may be blameless. He is also a smart guy and, after a bit more of a learning curve than Mike would require, Ernie is likely to develop into a good plant manager as well. But you get a lot more patience while you move up the learning curve if the broad base of support is already in place. Clearly, there are some lessons to be learned here when we get into the roles of the other staff managers. Each of them should find ways to become more visible and to begin to develop some grassroots support.

Suffice it to say, the manufacturing manager must have a presence on the shop floor. This should include making the daily rounds to stay in touch with supervisors, being visible and approachable by the hourly people, and seeing first-hand what kind of day the shop floor is having. My experience is that in a manager's zeal to be visible (the annoying term *managing by wandering around* comes to mind) he often falls victim to touring for the sake of touring and doesn't really see what's going on. In fact, he is just wandering around. Every walk out onto the shop floor should be a purposeful walk, where observations about people's behaviors, machine problems, safety concerns, and inventory issues are not passed by. Whether you're just going to the coffee machine or touring a particular cell or department, please remember to make your walks purposeful and train your supervisors to do the same thing. For example, on the way from the office to see one of the value stream managers, you are so focused on going directly to a specific place that tunnel vision sets in. With your head down, you miss opportunities to make eye contact and nod to operators as you walk down the main aisle. You pass by a pallet that has not been placed squarely and is protruding slightly into the aisle. In another cell on the other side of the aisle, the first-in, first-out (FIFO) lane ahead of the test machine is overflowing. What have you silently communicated while walking down the main aisle? What opportunities have been missed to assist your team? Observant hourly associates might assume that you are not very friendly or approachable, that safety isn't that important to you and neither is 5S, and that there is a major interruption in flow that needs immediate attention but you just kept going to "a meeting," which was the most important thing in your mind.

A purposeful walk simply requires that your head is up and your eyes are scanning from side to side looking for evidence of safety issues, flow interruption, things that aren't in their place that require attention. Leave the office 10 minutes early for your appointment so you can collect the appropriate supervisor along the way to be sure he's aware of and correcting these kinds of things. Or be late to the meeting. Addressing these shop floor issues is more important. That sends a powerful message to your silent observers as well.

As the manufacturing manager, you have the opportunity by nature of the position to become very knowledgeable in company and plant policies. Being accountable for so much of the total workforce demands that you are well informed enough to properly administer policies and to mentor and advise your subordinates. An absolute no-no is for the manufacturing manager to make a decision that is precedent setting and, in the worst case, becomes a de facto policy. Ignorance is not an excuse for this kind of an outcome. Know your plant's policies. If you don't know or are not sure, ask.

In the second half of the 1970s, our industry had a very strong resurgence of growth. Equipment utilization was at an all-time high. Constrained capacity needed to run every available hour of every week while capital projects with long lead times took shape and alternate routings and other debottlenecking projects were pursued. We had historically done a poor job with operator cross-training; thus, we were playing catch up so extruder operators could run more than just their normally assigned machine. For example, if an operator was running a 2.5-inch machine but the bottleneck was the 3.5-inch machines, then we wanted him to be flexible enough to help out with some overtime if a regular 3.5-inch machine operator needed a full weekend off.

Having been through a tough couple of years that included layoffs and short work weeks, many in the workforce loved getting the extra overtime. Some did not. But with lead times extending out as far as 52 weeks on "A" items, the plant was under enormous pressure to produce more, more, more. As a management team we had recognized how hard we were pushing certain machine operators, and in spite of our efforts we still needed them to work every Saturday and one or two Sundays each month. It was a rare exception when we asked for more than two Sundays a month.

During this time, a department manager who, because of the complaining of a number of his operators, decided to work his people only two Saturdays a month so they could have more time with their families. This sentiment was certainly easy to understand from a personal standpoint, but it was the wrong answer for customers until we could get additional capacity online. It was also the wrong answer because it was done unilaterally. Now the other three department managers who had their own problems with capacity constraints were having their operators come forward crying foul. They wanted the same deal. The human resources manager was not consulted. Without any other voice being involved, nobody had the opportunity to say, "Hey, wait a minute. If you do that in your area then what are the implications in the rest of the plant?" And, of course, as soon as the sales department heard about this the phones in the plant were ringing off the hook. Particularly annoying in this case is that now, in the eyes of his people, this department manager was their hero, and the rest of us were the bad guys. His action had undermined his peer group, his boss, the human resources manager, and the plant manager.

Needless to say, this was a precedent that had to very quickly be reversed with a series of department shift and staff meetings so that everyone understood why this scheduling practice could not be supported anywhere in the plant. The meetings took a lot of preparation and several hours to deliver until everyone had heard the same messages. Focus was on all of the actions being taken to lessen the burden on the same groups of operators every week. There was also a realistic projection of what the negative impact could be on our business if our lack of performance caused our customers to begin taking our business to competitors. In the end we managed to calm things down and help people understand that we needed their help to the maximum degree possible until we could work our way out of this happy problem of having too much business. We also made it clear that their supervisors would do everything humanly possible to help arrange coverage for a regular operator when he had something special coming up like a weekend birthday, anniversary, or visit from grandparents. Ultimately, we struck a compromise where the first and second shifts would work just 6 hours instead of 8 on Saturdays. This allowed the first-shift people to get off work at noon, and the second shift was finished at 6:00 p.m. That's the way the plant was scheduled for several years after until the plant went to a four-shift, 7-day operation. In the long run, one could argue that the end solution was a good one. But the precedent-setting process that made this an immediate management priority during a capacity shortage crisis was dead wrong.

No individual person, regardless of his position in the plant, has the authority to set policy. Plant leaders are responsible for administering policy. Policy changes or new policies must always be broadly vetted with the plant leadership team as well as any division or corporate leaders who also have a vested interest.

KEY RELATIONSHIPS WITH PEER GROUP

A key partnership in this regard is with the human resources manager and his staff. Whether it is knowing how the job posting system works or what the overtime policy is for hourly associates, the manufacturing manager is expected to have the right answers. In other areas such as medical benefits administration, the manufacturing manager should at the very least know the name and the phone number of the person in human resources who can provide the correct answer to an employee's questions. During the course of all of the daily interactions, the manufacturing manager becomes highly educated about plant policies, which gives him a major leg up on preparing for the plant manager's chair.

The manufacturing manager will also learn a lot about how the scheduling systems work and how his organization can become a proactive partner in meeting customer delivery promises. He must work closely with the materials manager and that team to understand his role. The scheduling function sometimes has to intervene and change the priorities of shop orders, most often to accommodate a customer's request. I think it is terribly important that the manufacturing manager gets to know scheduling personnel well enough to develop a real confidence in how schedules are managed.

Similarly, the manufacturing manager will learn the basics of the quality management systems and his critical role in setting the proper expectations with regard to International Organization for Standardization (ISO) 9000 or other quality system standards. ISO standards require comprehensive documentation of exactly what members of the plant team do under various circumstances. For example, what is the process when an operator knows he has just produced a part that is not per specification? What form is used? Who completes it? What is the process for dispositioning the rejected material? Who has the final say in this decision? What records are kept and for how long?

When the auditor checks compliance, the plant must have all of the key process steps well documented, and every person who uses that process must be performing it exactly as it is documented in the standard. Otherwise, it's a violation. These transactions require serious attention to detail and disciplined compliance with the standard work.

Some of the people involved in this process may work for the quality manager. Others likely report to the manufacturing supervisor. But let's be clear: It is the manufacturing manager's responsibility to make certain that each person in his organization understands that there is no compromise when it comes to quality standards. Each employee within the manufacturing chain of command is responsible for the quality he produces and for completing the necessary quality processes (e.g., testing, documentation, assisting with problem resolution) and, most importantly, for helping to ensure that no quality escapes get out the door to the customer. All quality issues are to be internalized and made totally invisible to customers (e.g., quality problems corrected, delivery date still met). The tone for all of this is set by the manufacturing manager.

Nothing infuriated me more as a leader in manufacturing than to have a quality person, a salesperson, engineer, whoever, suggest that he didn't trust the manufacturing organization to do the right thing on quality. I'm sure we've all heard about, perhaps even seen a few times, a manufacturing person abuse his authority and make a marginal call on product quality and ship it to the customer over the protests of a quality professional. This is simply unacceptable behavior. It is grounds for termination. The plant manager and manufacturing manager need to understand that and set the example by the decisions they make and the way they participate with the quality, engineering, and sales professionals on these kinds of issues. They must always be looking for the right answer from a customer and shareholder point of view.

That's not to say that there isn't good reason to have a debate from time to time, involving the customer in the decision, when the product's "form, fit and function" are adequate for the particular job in question. One particular example that comes to mind in the cable industry is on product color of insulated single conductors. Since colors are used for identification purposes, typically for termination, a blue wire being slightly lighter or darker than specified may not be critical to the customer. We shouldn't throw this product in the scrap dumpster without first vetting the question. On the other hand, if the cable does not meet the specified electrical performance then we're likely looking at a remake of the order because the cable will not function as advertised or specified. In any event, the senior quality executive breaks all the ties on these kinds of issues.

What about some of the other functions? How should the manufacturing manager deal with engineering, maintenance, safety, and environmental professionals? How should he set the tone with the manufacturing supervisors on how to interact with these groups? Let's start with engineering.

When I say engineering I use *process engineering* as a generic term for what others may call manufacturing engineers, Lean engineers, or industrial engineers. I prefer this term because *process* can mean the capital equipment that has been deployed or the process of data collection at the work center or in the cell, or it could mean the process map that describes the flow and the line balance of the cell. You get the idea. Since all work is a process, it's about having a team of professionals who are well trained in the basics of process management. It's a combination of mechanical engineering and industrial engineering with a generous helping of Lean and Six Sigma tools training stirred into the mix. These are professional problem solvers and the supervisor's best friend.

The biggest complaint I consistently heard from engineers during my career was how manufacturing often wasted their time. When I sought clarification of this comment it invariably came back as, "The supervisor, Fred, called me out to the extruder because the operator was having trouble getting the machine to run at rated speed. So, I dropped the productivity project I was mapping out for a Kaizen event and went directly to the machine. As I walked the line and reviewed the control panel with the operator, I found that the operator did not have the heat profiles set according to the setup instructions. He'd forgotten to change them when he did his last changeover of jobs. As soon as these corrections were made, the compound began to flow better and the machine was brought up to the rated speed. I feel like I wasted 30 minutes of my time solving a problem that the operator and supervisor should have resolved. In fact, it should have been resolved within a few minutes of the startup of the new job. It's frustrating that they think their time is worth more than mine when I have all these important projects I'm supposed to manage to improve the business."

How can anyone argue with the points made by this frustrated engineer? That's why it's so important for the manufacturing manager to weigh in early with his staff about his expectations as it relates to engineering. For example, he might clarify it in one of his first meetings with all the supervisors:

> I want to be sure we're all on the same page as it relates to how we should use the scarce resource of process engineering. As you know, their time is extremely valuable. We have only five of these technical support people in the whole plant, and they are the principal managers of productivity improvement projects, Kaizen events, and capital projects. We need their work to result in improved processes that will help to keep our plant competitive. That said, they are also here to support our needs when we bump into problems that are technical in nature and that we cannot solve ourselves after reasonable time and effort to do so. But here is a pledge that I expect each of you to honor every day: As the supervisor responsible for this cell, I will not call engineering for help until I have spoken directly with the operator and have reviewed the machine setup to be sure it is exactly per the operating instructions, until I have reviewed the materials being used to be sure they are the correct ones, until I have walked the wire line to check for any evidence of a machine maintenance issue. Only after the operator and I have done everything we are capable of doing will I call engineering for help. I would suggest the same kind of consideration for maintenance personnel.

If a nut is loose, the operator is usually capable of tightening it rather than bothering a skilled mechanic who gets paid to do the more difficult tasks that require his training and skill set. If the operator turned on the incorrect valve we don't need a maintenance person to fix that either. If manufacturing people are sensitive to these kinds of things with maintenance folks then it's reasonable to expect maintenance to come running when there is a legitimate need for their services to get a key machine back into production.

Another frustration I've often heard from maintenance types is that they aren't allowed to perform scheduled maintenance because the manufacturing supervisor says, "I cannot afford to let you have the machine this week. I need it for production." Then invariably, a week's delay turns into a month, and then all of a sudden there is an emergency breakdown. Now it's a crisis, and maintenance people are expected to stand on their heads to get the machine fixed—and, of course, it is always at a higher cost and always takes much longer to repair than if they'd been allowed to have the machine when it was scheduled for preventive maintenance (PM). The manufacturing manager, again, must weigh in early about how he feels about maintenance support and what his expectations are for manufacturing making the equipment available according to the PM schedule. It's a great example of "Pay me now or pay me more later" if manufacturing's collective heads aren't on straight on the topic of maintenance. Here again, the maintenance manager is a key ally in maintaining and improving shop floor reliability.

Another area I want to briefly touch on here is the role of the environmental, health, and safety person. Regardless of where in the organization this technical expert reports, as the manufacturing manager I want him to know right away to keep me totally abreast of any issues or concerns he has regarding environmental stewardship of the plant and for the safety of all employees. In the case of environmental issues, the manufacturing manager should be aware of all of the important projects that are planned for the year and to what extent his people will be involved in their implementation. In concert with the scheduling function, it's important that production downtime, if any, has been included in the plan for the plant so that no conflicts arise when the time comes. For example, a weekend shutdown may be required the second weekend of June to install a new air venting system on equipment to reduce airborne emissions. Or the water may need to be off for the weekend because of the need to repair leaks in piping that are allowing material contaminants into a storm sewer. The manufacturing manager must make it clear that he takes his responsibility seriously for doing the right thing as it pertains to all environmental issues. Scheduling accommodations simply must be made for these kinds of situations.

On the other hand, the environmental engineer, as the technical expert, must also understand and be sensitive to the costs of the downtime and the potential disruption to customer schedules. The manufacturing manager will likely want an early review of the project management plan and have the opportunity to be sure the engineer has done sufficient contingency planning to ensure that the downtime will be no worse than planned. But the tone on approved environmental projects should always be one of support to get the job done and done right.

Further, the manufacturing manager should be sponsoring projects to help the environmental engineer get out of the remediation business of solving the issues at the end of the pipe and begin moving toward prevention of the things that cause environmental issues at the front of the pipe. There is enormous synergy in this type of collaboration. Getting help on changing tool cleaner, for example, to one that is more environmentally friendly is a great example of the many things that can be done when both organizations are working together. My experience is that it also often results in cost reduction.

Safety is a staff function but a line responsibility. The safety engineer or coordinator should be the programmatic support for a robust safety program on the shop floor. By programmatic I mean the one who delivers training on basic safety programs such as the DuPont S.T.O.P. (Safety Training Observation Program) or forklift driver's training. He also is the one who might chair a plant safety committee, assist with the investigation of recordable accidents or near misses, and help with safety Kaizen events. In other words, he is a vocal and visible presence with a passion for having a safe workplace for all.

The manufacturing manager takes accountability for the safety and well-being of every employee under his umbrella. Again, that's the vast majority of people in a manufacturing plant. He sets the tone for the way people think and work. He personally follows up on safety investigations and corrective action. He sets the expectation; for example, "It is our job to ensure a safe working environment for every person who enters the plant and that results in zero accidents."

In the manufacturing cell or department, the supervisor leads in helping his people have the correct mind-set about safety, the right training, and active involvement in keeping their work areas safe. Every time the operator or supervisor walks the line, his eyes are actively engaged in looking for unsafe acts and conditions. When such acts or conditions are discovered they are dealt with immediately. The supervisor quickly learns that if he walks past a safety issue then he has just lost credibility with his workers regarding his priority on safety. Once again, walking silently by a safety issue communicates loudly that it's really not that important. One of the most obvious examples I've seen happen numerous times is when a forklift driver delivers material to an unauthorized storage location—in the worst case, in an aisle of the plant. This is never OK. But all the manufacturing manager has to do is to walk by one such occurrence without dealing with it, and he has just said through his inaction that leaving material in an aisle isn't that big of a deal. Wrong—it is a big deal. By the way, the supervisor may have stepped away from his area for a few minutes, so instead of the manufacturing manager covering for him by dealing positively with this he has just undermined his own supervisor who has consistently walked the company line on this issue.

The next relationship I would like to note is with the person who is responsible for computer systems in the plant. Many plants aren't large enough to have their own information technology (IT) staff but frequently have someone reporting to a staff manager who handles systems in-house with support from a division or corporate IT person. Often this position reports to the finance manager since many of the systems are to run financial reports, payroll, and plant metrics. I can't tell you how many times over my career that I've seen frustrated staff managers suffering with the "garbage out" they were trying to cope with from local systems. Most often, the reason for their frustration was because of lots of "garbage in" to the systems because hourly employees, nonexempt salaried clerical staff, or supervisors were not using the formal system correctly, if at all.

Here again, it's the manufacturing manager's job to set the expectation that his entire organization, without exception, will be trained and will use the formal reporting systems in a disciplined way. Examples of transactions that may be done by keying or by scanning bar code are reporting of quality rejections, materials scrapped, labor time and attendance, material transfers, and variations to standard machine speeds. If the plant manager and the staff are to have accurate reports with which to understand the plant's performance, then the manufacturing manager must elevate his expectations and champion accurate data input.

The final staff relationship I want to mention is between the manufacturing manager and the finance manager. It is fairly common for the finance manager (or plant controller) to have a dotted line relationship to the plant manager and a hard line report up the chain ultimately to the chief financial officer (CFO; more on that in Chapter 21). But on a day-to-day basis the finance manager works as an integral part of the plant manager's staff. He functions as a peer with the manufacturing manager.

As such, the manufacturing manager must respect the finance manager's responsibility to keep the plant manager fully apprised on issues that affect plant performance both good and bad. That said, the manufacturing manager should develop a solid working relationship such that the finance manager will openly share observations and concerns immediately. The manufacturing manager must make it clear that he wants timely and candid input so he can get in front of problems and opportunities. Early collaboration may get issues proactively resolved in a way that allows the performance forecast to be met or even improved upon. Worst case, early involvement allows the manufacturing manager to be the one to advise the plant manager of lurking issues that may impact the performance to forecast.

Ideally, the finance manager will be highlighting something the manufacturing manager already knows and is addressing. But the manufacturing manager always wants to be grateful (and to say so) to the team member who brings such issues to his attention. No killing the messenger.

The manufacturing manager also has to be sensitive to the fact that an issue that has lingered without resolution and puts plant performance at serious risk will be taken to the plant manager directly so there are no surprises. The finance manager would be well served to give the manufacturing manager a courtesy heads-up call, but ultimately the finance manager must make a judgment consistent with his fiduciary responsibilities. In some cases the finance manager would be irresponsible should he delay communicating with the plant manager. It is imperative that the plant manager gets a courtesy call before the plant finance manager elevates the issue to division or corporate finance (more on this in Chapter 21).

As you can see, supporting the plant manager, the manufacturing manager has a big job. On behalf of the plant manager, he is leading the shop floor efforts:

- As the champion for achieving and sustaining Stage 4 manufacturing excellence
- To ensure a safe working environment
- To deliver customer orders on time
- To meet the product quality expectations of the customer
- To execute the financial plans (make the numbers) that keep shareholders happy and earning a fair return on their investment
- To manage in a way that earns large numbers of employees' respect and hard work

He also sets the expectations for and ensures:

- The proper use of authorized formal systems
- The use of data-driven improvements (not seat-of-the-pants reactions)
- The correct skill sets and styles are resident in all supervisors
- The generous use of Lean and Six Sigma tools to improve the business every day
- The development of hourly associates
- That he is a visible and vocal presence as the shop floor leader

Many people regard execution as detail work that's beneath the dignity of a business leader. That's wrong. To the contrary, it's a leader's most important job.

Larry Bossidy, former chair, Honeywell International

We carry our wounded but we shoot the stragglers.

Doug Hitchon, former plant manager, Bundy Automotive

16

The Materials Manager's Role

This chapter is broadly titled to account for the fact that there are multiple functions represented by this position and that manufacturing plants may have all or just a part of the total scope implied here. As a result, I will address the key functions that I've most often seen in the plant over which the plant manager has a great degree of control. One important function, purchasing, often is not part of the plant organization but can obviously have profound impact on plant performance. In this case, close working relationships and right-headed division or corporate staff are required to help optimize business performance and customer service. But let's start with one of the classic plant responsibilities: production control, that is, capacity planning and the scheduling of customer deliveries.

PRODUCTION CONTROL OFFICE (PCO)

This function is critically important to the success of the plant and is a great training ground for up-and-comers because it presents a unique perspective on the plant operation. Typically, someone within PCO will serve as the link between the customer and the shop floor. This "outside-in" perspective is terribly important to help keep the blinders off with manufacturing supervision, engineers, and others who sometimes are too close to the day-to-day issues of the shop floor. It's easy for them to be so internally focused that the priority to serve customers slips down the list of things to do well. In this regard, as a plant manager, I always expected the materials manager to be the staff's customer advocate and to be sure the customer's interest is strongly represented around the staff table. Of course, excellent, seasoned staff will all think like that.

On the other hand, it can also be a slippery slope for the PCO person to remember to not make decisions on behalf of one customer, thereby putting other customers at risk. Sometimes just oiling the squeaky wheel is the wrong answer. In other words, the PCO person must work to optimize the outcome so that the customer in distress can win, the plant can win, and other customers don't suffer. It takes a savvy person to fill this role—one who is committed to customer service but who also understands the factory and its constraints. The promise for delivery must be executable and must include commitment on behalf of the shop floor organization. During my 25 years at Belden Wire & Cable, I had the pleasure of working with Ron McNew, who was the best I've ever seen at walking this tightrope. He knew the factory so well that he understood what it could and couldn't do from a scheduling perspective. In my experience, the only thing worse than missing a customer commitment is telling a customer the delivery date he wants to hear and then missing it. Ron was wise enough to know that telling the customer what he wanted to hear did not make such a date doable in the shop. He always got customers the best date he thought could be delivered as promised.

This all points to one of the primary roles the PCO group plays: scheduling the plant. Now before all the Lean zealots cry foul, I hasten to add that I'm talking about the planned schedule, not the shop floor execution of it. It remains critical that the PCO organization participate proactively in the formal sales and operations planning (S&OP) function with sales and product managers, manufacturing, engineering, and finance to plan materials and capacity such that constrained capacity doesn't get overcommitted. The PCO manager is normally the loudest voice on behalf of the plant at these formal planning sessions, which are typically managed to cover anything from a rolling 3-month period up to a full year. In my opinion, this functional leader requires specialized training to be the thought leader for his position. Some of the best training is accomplished by APICS (formerly American Production & Inventory Control Society—now known as the Association for Operations Management). The Certified in Production and Inventory Management (CPIM) is intended for production control professionals and is highly recommended.

The outcome of the S&OP is in effect a contract confirming that the sales team has agreed to both the forecast volume and the approximate mix of products that will be sold. Manufacturing, materials, and human resources management all weigh in to ensure that the correct manning and shift configuration are in place, that the right equipment is scheduled, and that the proper materials will be available to execute the plan. The product (marketing) manager projects the margin to be generated from this sales mix, and the engineering organization makes sure the technical support is in place. Finally, the finance manager rolls up the numbers, and the financial outcome of executing the plan is distributed to all senior management for review and approval. This is the typical process in a well-run company that sells a significant percentage of its products through distribution channels. For make-to-order (MTO) businesses and for the more sophisticated make-to-stock businesses, forecasting is less important since the production schedule is tied to actual incoming orders. There is no better forecast than a hard order. That said, if the MTO requires a special raw material that is not kept in inventory, then the lead time on that purchased item may well be the determining factor in what delivery promise will be confirmed to the customer. Even in this enlightened environment, it is likely that common raw materials will continue to be ordered using a basic reorder point system driven by aggregate known and forecasted demand.

I dwell on the planning systems here with purpose. In my experience one of the real misnomers about companies going to Lean manufacturing is that you no longer need the formal planning systems. To this I say a resounding "Bologna." Using your formal manufacturing resource planning (MRPII) systems, integrated with your engineering and cost systems, is still very much necessary and important.

Once the plan is in place, basic schedules of finished product delivery commitments are released to the shop floor based on the cycle time of the orders. The manufacturing organization then takes responsibility for executing the delivery promise to the customer. As cycle times are reduced on the shop floor, the plant can transition to a pull replenishment system. The evolution clearly is to use visual signals in place of computer systems.

For example, when a picking front in the warehouse becomes empty, a pallet from inventory is used to replenish the picking front. The removal of the pallet from inventory then triggers a card to the shop floor. The shop floor responds by replenishing the pallet to warehouse storage. This is a major step to respond closer to real demand as opposed to building to forecast. The only thing better than this is to have your major distributors set up to send the plant an electronic card. This allows the plant to respond directly to customer demand and bypasses the company warehouse step all together—at least on high-volume "A" inventory items.

The PCO organization will be intimately involved along with manufacturing, engineering, and the customer in designing the appropriate shop floor execution system. It is well beyond our scope here to get into the details, but suffice it to say that many visual Lean tools such as kanbans, first-in, first-out (FIFO) lanes, and min–max replenishment signals will be integrated into the shop floor control system. It is important that members of the PCO staff are visible and involved in the design and execution on the shop floor.

Purchasing or Sourcing

Another critical responsibility under the umbrella of materials management is the purchase of the raw materials and supplies necessary to execute the production plan. Sometimes this responsibility is resident in the plant. Often it is not, especially in companies that have a lot of common raw materials and multiple factories to support. Because of this I shall address this function assuming that purchasing is done centrally to take full advantage of corporate synergies.

Where the purchasing function is performed doesn't change what must be done. Unfortunately, however, when purchasing is done at a division or corporate office, plant personnel sometimes don't get the level of communication, understanding, and attention they deserve. All too frequently, division or corporate sourcing personnel forget why they have their jobs: to support taking great care of paying customers by ensuring that the right materials, in the right quantities, at the right quality level are in the right place at the right time and at a competitive cost.

Too often the division or corporate managers and buyers are too focused on delivering favorable variances by getting lower costs on materials than are in the standard cost for the year. The more important point is to make sure that the materials are manufacturable in the plant without adverse consequences. For example, if purchasing gets material delivered with a $.05 per pound price decrease, it's only a savings if the material meets the same raw material specification as before, if the material yields are the same, if machine speeds are maintained, and if scrap and rework levels aren't increased. It is a "total value delivered" mind-set that is important. Hopefully the mind-set at a division or corporate office will be in alignment here. If not then the plant materials manager and possibly the plant manager will have to intervene and help to cause different and more supportive behaviors for running the business. Of course, if purchasing is done at the plant level, then the plant manager would expect nothing less from his own materials manager and sourcing team.

At the plant level, in concert with plant engineering and purchasing, the materials manager is expected to be proactive in making certain that the specifications for raw materials are accurate and are being properly maintained for each material and supplier. As an aside, I always call them suppliers. I believe it's a much more professional and respectful moniker as opposed to vendor. A supplier is an extension of my business—a collaborative, important, strategic partner for our mutual success. A vendor is a person who sells peanuts at the ballgame.

There should also be an ongoing joint project with engineering, purchasing, and the suppliers themselves aimed at raw materials commonality. This is a win–win–win–win–win. Customers are more likely to get their orders delivered on time because of the elimination of unnecessary complexity in manufacturing the item. Company margins are likely to improve slightly due to a lower cost part because of its higher production volume. Process engineers have to engineer fewer processes and to maintain fewer records in the specification files. Materials and manufacturing people have to keep track of far fewer items in inventory and on Bills of Materials on production orders. Purchasing has fewer materials to negotiate and on which to maintain contracts. Suppliers have fewer parts to manage and longer production runs on the common parts they continue to supply. All must, however, keep an eye on those pesky product (design) engineers who must then engineer specialty products from a group of common parts wherever possible. Pulling in a new supplier or a new part when it is not necessary is counterproductive regardless of how much fun it may be for the creative nature of the design engineer. So an audit of new specifications for this item must be added to the regular new products audit done by process engineering. Don't optimize commonality only to begin losing ground immediately with the next design that comes off the drawing board. This is an important mind-set change that must occur in product engineering to be fully supportive of manufacturing excellence!

Purchasing professionals should also strive to reduce the number of suppliers to the theoretical minimum number required to successfully manage the business. My experience is that traditional plants have four to five times the number of suppliers required to run the business. Typically this is the result of many years of adding suppliers for a variety of reasons, often good ones, but without enough emphasis being placed on eliminating those that are no longer strategically important.

Combine the lack of good commonality discipline in design engineering and purchasing with a lack of a comprehensive sourcing strategy, and you end up with a much larger list of suppliers than is necessary. It's also a much longer list to maintain and manage. It's unnecessarily expensive and inefficient and makes no sense to the business. From the supplier's perspective, it's an opportunity to improve revenues and earnings by being one of the preferred suppliers of your company.

In the early 1990s the purchasing manager, Mike Foley, and I were the impetus for creating a supplier advisory council (SAC) for the Electronic Division at Belden Wire & Cable. We were fortunate to enlist the help of Ken Stork. Ken was a foremost authority at the time on supplier management based on the extraordinary, breakthrough kind of work he did at Motorola to develop world-class suppliers. He helped us understand that companies who were seeking manufacturing excellence had to also have world-class suppliers to support them. He also helped us to understand and outline what the basic expectations should be for a supplier advisory council that would really add value to our journey.

Once we had our spec developed, we selected the top executive at a supplier company to represent his commodity group (e.g., metals, compounds, packaging) for a 2-year term. Six such executives were rotated out on a schedule of two per year. That assured good continuity over time. The charter members of the SAC were invaluable to establishing the mission and the ground rules for our interactions and also were very supportive and flexible in extending their terms on the council allowing us to set up the staggered rotation. The one subject that was always off-limits for discussion was anything that had to do with supplier costs and pricing. We also asked that each commodity group representative wear the big hat and try to represent his group of suppliers rather than just to represent his own company.

One of their most important inputs to us went something like this: "You have high expectations and tough metrics for your suppliers. We don't have a problem with that because you cause our companies to get better to satisfy your requirements. So we benefit as do our other customers. What we do have a problem with, however, is that we make the investments in our systems, people, equipment on your behalf, and then you continue to split the business among three or four other suppliers, including some who aren't making the investments and are not performing up to your standards. Where's the quid pro quo that recognizes our investments and rewards our being one of your top-performing suppliers?"

This was great input and convinced us that we should proceed with an objective of reducing our supply base by 80% in 5 years. We called large groups of our suppliers into an auditorium and announced the reasons for our supplier reduction program: why it was important to them and us; the standard metrics and the minimum hurdles that all would be required to meet. We announced that the best performer in each commodity group, with equal emphasis on quality, service, and delivered cost, would receive 70% to 80% of our business. This supplier would be so reliable that there would be no requirement for incoming inspection of their incoming materials. Their materials would be delivered directly to the line or into our inventory. This company would be a "certified supplier" after having passed a comprehensive audit. So suppliers wouldn't sit on their laurels, the audit was repeated on an annual basis.

The second-best supplier in each group would receive the remaining 20% to 30% of our business. Other suppliers in that group would lose us as a customer. The only exception to this general rule in my experience is where common sense dictates otherwise. Some materials are so critical to the business that an uninterrupted supply of it is essential to the viability of the business. In wire and cable that was copper. Typically the copper buy was spread among three major suppliers to ensure adequate capacity was available to service our needs in the event one of the three had production problems or there was an unexpected surge in demand.

The improvement within the much smaller supplier base over the next 5 years was nothing short of outstanding. Suppliers that up until then had been constraints to progress on the shop floor were now part of the solution instead of being part of the problem. Further, our certified supplier partners were now being pulled in early with our technical people to collaboratively work on new materials or new product development. These suppliers became an extension of our own management team—another win–win proposition.

In summary, this is all about creating the proper level of commitment and accountability from your supplier base. So whether the accountability for the purchasing function resides in the plant, the excellent plant manager expects his materials manager to be on the point and driving the aforementioned initiatives to create and then to sustain: commonality of raw materials, accuracy of raw materials specifications, significantly reducing the number of suppliers and dealing with them as important strategic partners. This process is likely to take 3–5 years to fully implement, but it is critical staff work to support manufacturing excellence long-term.

Formal Systems Zealot

I mentioned earlier the need for the continued and disciplined use of formal systems whether you're in a traditional plant or a Lean and Six Sigma plant. Some of you may be thinking, "Why do I still need a bunch of formal systems now that I'm going Lean?" Well, let me answer your question with a series of my own questions:

- Is it still important to have accurate Bills of Materials and routers?
- Is it still important to have accurate engineering information relative to the correct machine speeds as well as setup information?
- Is it still important to have accurate production standards with which to measure and plan machine capacity?
- Is it still important for the material usage figures in the specifications to be accurate, such as the pounds of plastic compound being used per thousand feet of cable being produced?
- Is it still important to have a good handle on product costs?
- Is it still important for sales and product management to understand what their margins are at current pricing?
- Is it still important that data entry from the shop floor is both accurate and timely?
- Is it still important that there is good engineering control on all specification changes?
- Is it still important that raw material specs are accurate and that the supplier always has the correct revision number?
- Have I made my case by now?

Because of the intimate involvement in the planning system on behalf of customers, as a plant manager I always expect the materials manager to be the zealot on formal planning systems with his peer group—to be the conscience of the plant on this topic. As we'll see in later chapters, there is also a key role with the disciplined use of formal systems for the plant controller and finance manager.

Lean Zealot

The plant manager counts on his materials management team to be zealots in the quest to reduce cycle times on the shop floor, in the offices, and at supplier plants. In my experience I've seen sourcing people beating up suppliers for lead-time reductions when far bigger cycle time issues in their own plant weren't being addressed. I've also seen plants that have done a great job driving down their internal cycle times to the point where raw material availability (i.e., the supplier's lead time) had become the constraint for the plant quoting new business. Let's be clear: the terms *lead time* and *cycle time* are not synonymous. I've been especially surprised to find materials managers who did not understand the difference between lead time and cycle time. Let me share my own definition of the terms.

Lead time is the total elapsed time from when the customer places the order until the order delivers complete at the customer's dock. Lead time looks something like this:

> [customer places order → order entry cycle → conversion to manufacturing order cycle time → supplier lead time for materials → total manufacturing/assembly cycle time → packaging/shipping cycle time → transit time to customer]

In other words, lead time is the sum total of all of the individual cycle times involved in the process from order entry to shipping. Note in the previous example that customer lead time is the sum total of seven separate cycle times. Lead times, therefore, will fluctuate depending on how well the process inputs and outputs are in balance. It is always a question of *capacity*.

For example, let's assume that order entry is expected to process and order in 1 day. However, due to a significant increase in order activity it is now taking 4 days to get an order through the process. If everything else downstream is executing per their normal cycle times, then the total lead time to customer would be extended by 3 days (i.e., 4 days instead of 1 day cycle time through the order entry process). Again, it's a question of capacity for the order processing function.

The short-term answer to avoid extending the lead time being quoted might include overtime for the order entry people or perhaps moving cross-trained associates into the department for a temporary period to help. If the order surge persists and is expected to continue indefinitely then a more permanent solution may be necessary, such as substantially improve the current process to create new capacity or, in the worst case, hire additional people to prevent increasing the lead time to customers.

Does this sound familiar? It's the same kind of action that is required in supplier capacity for materials or for adjusting capacity in your plant when machine capacity is the cause of extending lead times. Every possible action should be anticipated and executed to avoid increasing lead times to customers. If your competitors are better than you are at making these kinds of capacity changes then they will be taking bites out of your company's market share. The materials manager must be the conscience of the entire plant to anticipate and lead these responses. This is another clear case where this key member of staff must assert himself as the customer zealot on behalf of the plant.

Parallel to this effort is creating the awareness through this value stream to be constantly looking for ways to reduce the cycle time of each segment of the total process. These actions will eventually have sufficient cumulative effect to compress the lead time and may well result in creating a competitive advantage. When there is a concern about lead times stretching out, the fingers from around the company tend to point at manufacturing as the problem. Sometimes manufacturing is to blame and sometimes it is not. I have been in situations where the manufacturing cycle time was actually less than the time it took to enter the order and provide the shop packet of job instructions to the shop floor. In this case, working harder to reduce the manufacturing cycle time would have been a misdirected waste of time—pure *muda*. In this particular case the priority was to map the front office process and significantly improve it. Fortunately, all work is a process and can be mapped. And if it can be mapped then it can be improved. The really good news is that the same process that works on the shop floor also works in the office.

My favorite tools for solving this kind of a problem are the value stream map (or process map if you prefer) and the manufacturing cycle efficiency (MCE) index. Take the process in question, and map it completely from start to finish. Once that's done the individual tasks and the time it takes to perform each one are sorted and applied to this simple formula:

- MCE = value-added time/total elapsed time
- Value-added time: Time the machine is running quality product, at the rated speed in the proper quantity

- Nonvalue-added time: Time to make setups; time running material not to specification; time spent running overproduction; time spent moving materials or looking for materials; time spent on maintenance downtime, etc.

In the office this could be, for example, that part of the order entry process that is critical to releasing the information required to get the manufacturing process started as soon as possible. Moving paper from one inbox to another would be nonvalue-added, as would any queue time.

Over the years I've seen it proven time and again that an MCE of >.50 is excellent, whereas an MCE of .05 is more typical and represents a significant improvement opportunity. Teach this thinking and use these tools in the office to capture the benefits that reside there. The MCE is a great tool to identify and to focus effort on improving the constraint regardless of where the constraint is in the overall process. This is simply a great way to take Lean thinking and Lean tools from the shop floor into the office—and the materials manager should be leading the way on behalf of the plant manager.

In summary, the plant manager needs and expects the materials manager to play a major role in the success of a manufacturing plant. In short, the materials manager's role is to deliver materials management excellence over the entire scope of his job every day. It's as simple and as difficult as that.

> Bottlenecks are only a good thing if you manufacture bottles.
>
> **As seen in a J.D. Edwards ad**

> It is not necessary to change. Survival is not mandatory.
>
> **W. Edwards Deming, world-renowned quality guru**

17

The Process Engineering Manager's Role

The plant manager who is creating a continuous improvement culture needs staff managers who are helping to lead the revolution. Perhaps no person is more critical in this endeavor over the long haul than the process engineering manager (PEM).

THE ENGINEERING STAFF

As the primary technical support in the plant, the plant manager expects the PEM to be the primary owner of continuous productivity improvements on the shop floor. Ideally we're looking for a leader who recruits "shop rats" who have insatiable curiosity and are, instinctively, always looking for a better way to eliminate waste and improve processes. Engineers who aren't genuinely interested in getting their hands dirty or in coming in on the off shifts to help improve the plant have the wrong DNA for what is required to deliver manufacturing excellence.

The kind of engineers I always wanted to have were those who were constantly walking the wire line looking for ways to do things better or looking for things that didn't quite look or sound right so they could stay in front of any issues with the process. The best engineers are always visible on the shop floor and have a great rapport with the machine operators. Why? Because they are constantly tapping the operator's expertise looking for ideas and input—showing him the respect that he deserved but often hadn't gotten over his career. In addition to a good working relationship with operators, the engineer must of course be technically competent.

Depending on what kind of manufacturing operation it is (e.g., compounding), you may need specific or specialized knowledge about plastic or rubber technology. Most often, however, degrees in mechanical, industrial, or manufacturing engineering provide the basic skills necessary coming right off the campus. In addition, some companies have Lean or Six Sigma Green or Black Belts who are highly skilled in the application of a broad array of Lean and Six Sigma tools. Black Belts, in fact, should be capable of teaching Green Belts and others, including hourly associates, who simply need basic tools such as fishbone diagramming, 5S, and SMED* (quick changeover) events. But if you need a regression analysis, a designed experiment, or a process capability study, any engineer in the department should be capable of delivering it.

My experience is that far too many engineering departments are woefully lacking in engineering staff skills necessary to excel. Unfortunately, it has also been a rare occasion during my 35 years in manufacturing for every member of a plant engineering staff to have the right kind of a leader in place as the manager. Even rarer has been having a staff of proactive, passionate engineers always aggressively on the move and driving process improvement to make the business better.

More common, unfortunately, is a group that doesn't have these skill sets, a group that doesn't volunteer to help out on the off shifts, a group that gets so tangled up in patching over shop floor problems instead of solving them that they are constantly harried and can never seem to find the time to get in front of it and really take control of their jobs and how they spend their time. They fail to understand and to believe that enabling manufacturing excellence is not about firefighting—it's about fire prevention.

One of the first things a new plant manager must do is to assess his staff and decide within the first 6 months who is part of the problem and who is part of the solution. That, of course, starts with the plant manager's direct reports. However, the next two most important places to evaluate are the manufacturing supervisors and process engineers. This is where much of the resistance (mostly passive) typically resides, and these associates very often become an obstacle to progress in the best case and a cancer in the worst case. How will we ever change culture and get hourly associate buy-in if these two groups aren't both competent in their jobs and passionate about improvements? We'll simply never get the hourly associate buy-in

---

* An acronym for single minute exchange of die, SMED is often used interchangeably with quick changeover. Technically, SMED literally means changeover times on machines that require less than 10 minutes from last good piece to first good piece.

and active participation if they see firsthand the hypocrisy between what their top management is saying and what their direct supervisor and their process engineer are thinking and doing.

CAPITAL APPROPRIATIONS

This is another important part of what the engineering staff brings to the plant. Frequently the PEM, in collaboration with peer staff managers, is responsible for recommending capital expenditure budgets and for implementing the resulting approved projects. Capital spending can be an important ingredient in manufacturing excellence. It can also become a colossal waste of company resources. Let me cite a specific example from my experience.

I once had a plant manager who sent me a capital expenditure request (CER) for over $250,000 for a 30-inch cabler that was "required" to debottleneck an electronic cable process. Mind you, this CER had already been approved by the following list of people in the plant before being sent to me: project engineer, manufacturing supervisor, engineering manager, manufacturing manager, production control manager, finance manager, and plant manager. Since the products made in that plant were oversold and the company was having a very good year in terms of cash flow from operations and net income, I was certainly open-minded about the need to spend some money to increase sales and earnings on these products. However, as I read through the CER I was looking for a critical piece of information: What was the overall equipment effectiveness (OEE) of the three cablers the plant already had? I found no reference to that, so I called the plant manager and asked the question. He didn't know. Further, the project engineer didn't know, and nobody who had signed off on the CER at the plant level had asked the question.

To make a long, sad story short, the OEE on the three cablers in operation averaged 45%. My experience was that this kind of equipment should have at least a 75% OEE. It doesn't take a math major to figure out that spending capital dollars increasing the labor costs and the depreciation burden of the plant were the wrong answers. When the plant manager and his engineering staff were challenged to improve OEE on three cablers by 30 points, they found two major issues. First, the supervisor and his operators didn't understand that the 30-inch cablers were the bottleneck, so they weren't diligent about making certain those machines never stopped during lunch time or shift changes. Cablers simply weren't recognized as "continuous operations," so always running them was an industry paradigm shift. In the past all that mattered was keeping extruders running. The second issue was that operators were spending way too much time making setups. "Gang changeovers"* and specific SMED projects that significantly reduced setup time were successful enough that the overall outcome from these two initiatives provided capacity in excess of what was required to debottleneck the process.

This example is why the engineering manager must think the right way and surround himself with staff members who also think the right way. In this particular case, the plant manager knew that the company was not in a constrained capital environment and, in fact, was spending tens of millions of dollars that year to expand capacity across a number of product lines in the business. As a result, he and his engineering, manufacturing, and finance team all either got lazy or simply stopped thinking and didn't even do the minimum level of due diligence on the need for such an investment. The moral to the story is that capital spending should be used as a last resort when all other noncapital options have been exhausted. As you might imagine, this kind of situation was just a symptom of the malaise that was pervasive in this plant at the time. Within the next 12 months there was a considerable shakeup of the management team in this plant. The correct mind-set was simply not in place.

That said, once capital investment is determined to be the correct answer, there are a number of other considerations. For example, is there a new-generation, state-of-the-art machine that we should consider?† Is it made by the same manufacturer or by another company where we have little or no experience? If we choose new technology, is the improvement sufficient enough that we should ignore the additional costs of having to train operators on a different process? What about the maintenance crew? Does it make sense to have to train them on a new machine? Does it make sense to stock new (additional) spare parts to accommodate the new manufacturer? Unfortunately, these kinds of issues are too often downplayed, if they were

* Other cell members help to make the setup while their machines are running or else on idle time from nonconstrained work centers.

† A word of caution here: You may not want to rush to be the first to buy the new technology. My experience is that early customers of equipment marketed as "state-of-the-art" end up being the plant laboratory for the manufacturer whose equipment still has development issues. That's not to say never be first; only to be sure that you go into it with both eyes wide open.

ever considered at all. An excellent PEM will be joined at the hip with the maintenance, manufacturing, and materials managers to ensure the correct decision is made for the plant. This cross-functional collaboration almost always results in an optimum outcome for the plant.

Once the technology decision has been made, it is the PEM's job to ensure that the CER process is thorough and complete and answers the questions that he and his team anticipate will be in the minds of the reviewers and approvers above the plant level. For example, how thoughtful is the support commentary for why the investment is important to the business and why should it be done now? How complete and accurate is the content of the CER? How comprehensive and clear is the performance specification for the equipment that will be sent to the supplier? This is critical to guarantee that the supplier has all the information necessary to manufacture the new machine so that it meets all the requirements of the buyer. Make sure your sourcing person is involved in this process.

One of the best practices I've observed over the years is that an outstanding plant manager would use his direct reports to vet the quality of the CER prior to submitting for division or corporate approvals. Not only did it establish a better quality CER, but it was also a great learning opportunity for the staff to think bigger—through the eyes of the executive team that had to ultimately make the decision on the investment being proposed. The ultimate test for me was always this: If this was my business and it was my name on the smokestack outside, would I spend this money? Our challenge as leaders is to get all of our people to think the same way when it comes to spending company money—whether for capital investments or for expense reports.

SYSTEMS INFRASTRUCTURE

The next major area is the important role the PEM and his people play in making sure that the infrastructure used to run a great plant is in place and is being properly maintained. As for the physical infrastructure (e.g., water, gas, electrical, air, steam) we'll leave that for the plant engineering and maintenance folks.

We expect process engineers to keep formal systems like bills of materials (BOMs), material usage factors, routers, and operating instructions accurate and up to date. Maintaining these important records in paper files or on Excel spreadsheets and the like simply isn't acceptable. What are the authorized, formal systems in which the company has invested millions of dollars? Are they being used for the purpose of having common, integrated databases so that all functions are working off the same data to run the business? What is the process for updating these systems when material costs change, when usage or scrap factors change, when machine speeds change, when setup instructions change? What is the formal system for making changes? How robust is the formal engineering change system? This disciplined use of authorized, formal systems is one of the first things I always look for when assessing a plant's readiness to seriously pursue manufacturing excellence. Frankly, there's little if any chance of sustaining success unless formal systems are nailed down tightly and all users are properly trained. The disciplined use of authorized, formal systems is one of the critical infrastructure items in manufacturing excellence, is a great example of standard work, and is a shared responsibility of leadership. Note that engineering, materials, and finance tend to be the major flag bearers on this one.

And finally, a note on standard work. When a plant is going Lean, one of the changes that will be noticed first on the shop floor is a heightened expectation of process discipline. For example, it is no longer OK for a machine setup to be a work of art. Standard work is required; that is, there is only one correct way to make a setup on a machine. That approved method will be documented, and operators will be trained and required to make setups exactly that way—no excuses, no exceptions. If a better way comes along, the standard work will be changed.

There will be a loud cry of how inflexible the leadership has become. Early on there will be lots of frustration. But leaders who stick to their guns on the basic expectations of a disciplined use of standard work will have started the manufacturing revolution in earnest and ultimately the necessary culture change. Critical here is that process engineering and manufacturing supervisors understand, buy into, and will sustain the higher expectations.

You may hear a crescendo of carping that "I can't try anything new anymore. All they want now are robots to do everything the same way every time." Of course this isn't completely true. We do want compliance to the requirements of standard work. It's the same discipline as for 5S and lots of other things associated with changing a culture. On the other hand, we also want associates to understand that we don't want them to stop thinking. If someone thinks he has a better way, his idea will be tested in a controlled trial. If the new way he has suggested is indeed an improvement then the new method will become the new standard work. Engineering will document the new process, all appropriate operators will be trained, and the new process will be followed all the time by everyone until a better way is suggested again, trialed, and implemented.

PROCESS ENGINEERING—A KEY TEAM MEMBER

Because the engineering leader will be the focal point for the technical side of managing materials, products, and processes in the plant, his selection is critical. He has knowledge and a skill set that doesn't reside anywhere else in the plant and, arguably, is the toughest set of skills to replace and maintain. Absent a strong engineering group, the deterioration of plant process control is simply a matter of time. Chemistry among staff members is always critical but is especially so in engineering. The engineering team members must understand that it is their role to deliver strong indirect support to the manufacturing team. It is absolutely imperative that they are aligned with the manufacturing priority set. Without a passionate understanding of this, friction and counterproductive behaviors will result that are costly to customers and shareholders. And fatal to sustaining manufacturing excellence.

> Simplicity is the ultimate sophistication.
>
> **Dave Garwood, author, *Bills of Material—Structured for Excellence***

> There is a better way. Find it.
>
> **Thomas Alva Edison, inventor and founder,**
> **General Electric (1847–1931)**

18

The Maintenance Manager's Role

As a new plant manager in 1980 I scheduled a one-on-one discussion with each member of my staff. In my first session with the maintenance manager I asked him to educate me on the state of the facility from his point of view. I also wanted to make it clear that he had my full support as the steward of our fixed assets. On my watch there would be no deterioration of the company's assets aside from normal wear and tear and depreciation on the equipment.

I was fortunate to have had great stewards of the facility before me who had set the example. For the times, the plant already had a serviceable preventive maintenance system. It was a paper card system, but it worked. One of my wise predecessors had implemented a robust rolling 3-year facility maintenance program that had religiously been continued over the years. Its purpose was to keep all real estate assets and plant services assets in the proper state of repair and replacement at all times. It also allowed the plant team to put upcoming major capital expenditure needs on the table with more senior leaders so there were no surprises later when the funds were requested. The really big numbers (i.e., over $1 million) were typically identified a year or more before the actual need to aid in the financial planning process. These are things I came to take for granted until I got into acquired plants a few years later and found out how progressive the maintenance systems had been in the Richmond, Indiana, plant of Belden Wire & Cable Company in the early 1970s. In fact, the 3-year rolling facility maintenance plan became a model that I applied in all plants for the rest of my career.

PREVENTIVE MAINTENANCE

In my view, this is the most important process that the maintenance manager must champion. It takes a disciplined manager, a passionate leader who understands his leadership role in keeping equipment in specified operating condition. It takes a team player who understands that his two primary roles are to make sure that key facility infrastructure systems such as electrical, gas, steam, and air are always available through quality maintenance and to make sure that production equipment has zero unplanned downtime, particularly on constraints.

Here's a manufacturing truism: If you run equipment to failure, it will always cost more (usually a lot more) and take longer to repair than if the equipment had been scheduled to be shut down for preventive maintenance. I've never understood the mentality of plant and maintenance leadership who choose to run equipment to failure, but I can't even count the number of facilities I've seen over the years that did just that.

Today many plants have computerized maintenance systems that provide the opportunity to accumulate and analyze failure data by machine, the number of hours the equipment operated between failures, the cost of each repair. Properly maintained databases are rich with information that can be used not only to fine-tune the preventive maintenance system but also to provide the framework to evolve into a predictive maintenance environment.

For example, let's say we've accumulated 3 years of data on equipment that is run continuously without ever stopping. Our data suggest that the bearings wear out and require replacement at approximately 10,800 hours of operation, assuming proper preventive maintenance and lubrication of the machine shafts and other related moving parts. Knowing from the unfortunate experience of having ruined a shaft due to seized-up bearings, we learned that it is very costly and time-consuming to make this kind of a major repair. Instead, we use the data to put in a scheduled bearing replacement on this kind of equipment at every 10,000 hours of operation. Machine by machine this kind of data analysis will result in the savings of huge dollars, hundreds of hours of downtime, and many missed customer promises.

Isn't this just like apple pie and the American flag? Isn't this a no-brainer to think about maintenance in these terms? Sadly, I must report usually not. In my experience maintenance typically is one of the worst managed functions in the plant. Of course we can likely all cite notable exceptions to this: in my case, Jim Norrick, absolutely the best I've ever worked with in leading maintenance with appropriately high expectations. But in the vast majority of the plants I've been in, maintenance has been one of the ripest targets for improvement. How can one legitimately strive for manufacturing excellence without having maintenance excellence in place? The short answer is you can't.

Some plants have undertaken total productive maintenance (TPM), which formalizes the path to maintenance excellence. I applaud these efforts with one caveat: Some plants have embarked on this initiative independently. My advice is that each functional staff leader should have a strategy to drive functional excellence. However, each must be integrated into a comprehensive plant strategy that results in manufacturing excellence. Isn't that the ultimate objective in a manufacturing plant?

The Rolling 3-Year Facility Maintenance Plan

As mentioned earlier, this formal tool is important to attaining and then sustaining excellence. The way I learned this was from one of my mentors who was the plant manager at the time I was the manufacturing manager. I had heard stories about how the plant manager, manufacturing manager, engineering manager, environmental health and safety (EH&S) manager, and maintenance manager got together one day each summer to do a comprehensive tour of the plant, inside and out. But this time, in summer 1979, I was to be a part of it.

The plant had two stories and covered about 800,000 square feet. The main structure was from the 1920s and had the unforgettable sawtooth roof design with lots of leaky windows on the vertical plane. Newer parts of the plant had rubber membrane roofs, asphalt and stone, and other modern marvels of the day. In any case it was several acres of roof to maintain at a very high expense. The basement contained the arteries of the plant's blood. The tunnels that ran under the plant contained water lines, air lines, gas lines—most of the plant services that were vital to the plant's operation.

Led by the maintenance manager, we tromped through the basement and into the tunnels following the piping that carried critical plant services. We also looked at everything from electrical load centers to aisles that needed to be restriped or damaged overhead doors that needed to be repaired.

After touring the basement and the main floor, we then took a walk around the entire acreage of the property. This is where the environmental health and safety manager was the most engaged. Were there any issues with storm drains? City sewer drains? Any loose pellets in the area where plastic was unloaded? Any asbestos or lead paint areas to remediate? Any driveways or dock areas that needed repair? Any dock doors that should be replaced? Any damage to outside fences or gates? Any compound silos that required painting? Any areas of the exterior building that needed repair or painting? Any repairs coming up on the rail spur leading into the plant?

Then it was up to walk the roof. Fortunately, the maintenance manager had done this enough times to know how to guide us on the tour so we could see what we needed to see without endangering ourselves. He always brought along a schematic of the roof that showed the construction date of each section of roof along with the last time maintenance had been performed on each section. Based on this information as well as records indicating where in the plant there had been leakage during heavy rains, we were able to establish the priority of where the money should be spent for the next 3 years of roof repairs. Since there was an annual review, unforeseen changes could be taken into account and the priorities reset as appropriate a year later. Maintaining this roof was a never-ending process. Each year's capital budget contained somewhere between $200,000 and $250,000 for roof repairs. On the roof we were also guided by the environmental health and safety manager with regard to which roof vents, for example, required permits as well as the current status of any renewals with the U.S. Environmental Protection Agency (EPA).

Once the list of items found on the tour was digested by the maintenance and plant engineering folks, the maintenance manager made his recommendation for projects and funding, by year, for the next 3 years. The Year 1 list reflected the number one priorities to be accomplished in the upcoming budget year. Years 2 and 3 were priority two and three, respectively. This report was due each year by the end of August so that it was ready when the capital budgeting process started in September. The maintenance manager understood that he didn't have a blank check so he'd have to make some priority calls. But he also knew that for critical items he would get the capital funds he needed.

For example, the inspection may have identified three compound silos that were on the list to be painted. However, one of those had been on the list now for the third year, and some early signs of rust were visible. That one had to be done in year 1 of the plan. The other two could be scheduled in future periods. We had visibility that they would need to be done in future periods but understood that funding was not yet required.

This is a good and necessary process and one that I required for all new plants that came into the company during my time at Belden. When I went to General Cable I installed the same process for all of those plants. To the best of my knowledge the process lives on in both companies. The question is do you have such a process in your plant?

Maintenance Staffing

The next key area for a maintenance manager is to be sure that a robust succession plan is in place for all skilled positions. Unlike most other hourly positions in the plant, the maintenance area has a team with very specialized skills that require a long learning curve. Some of the positions require a formal apprenticeship, which takes years to complete. Unfortunately I've known very few maintenance leaders over the years who had a calendarized plan for sustaining proper skill levels. Typically it went something like this: "Good morning, boss. I'm getting ready to do next year's budget, and Harry is going to retire at the end of April. I guess I need to start a pipe-fitter apprentice to take his place. Is that OK with you?"

After slapping his forehead in exasperation the boss says, "Yes, of course. But where's your plan to get out in front of this and manage it more proactively in the future? It's too important to the future of the plant for us not to have a formal plan that addresses maintenance attrition." Most businesses today have a formal performance evaluation process each year that includes succession planning for salaried people. Maintenance is the one area where the same thoughtfulness and forward planning should also be done for the skilled hourly positions.

Like all of the other staff functions, the plant manager expects the maintenance leader to think the right way and to use formal processes to manage the function. Maintenance is one of the most susceptible (maybe second only to manufacturing supervisors) to being among the plant's best firefighters. Most who have grown up in a traditional manufacturing culture earned their spurs by being the best firefighters and problem solvers. While I never want to discount years of knowledge and experience as an important input, using data to solve problems is necessary over the long-term to sustain excellence. I'm always reminded of the quote from world-renowned quality guru W. Edwards Deming: "In God we trust. All others bring data." Maintenance excellence requires outstanding fire prevention, not firefighting.

> Murphy's Law: If it can go wrong it will. Mistake-proofing corollary: If it can't go wrong, it won't.
>
> **Charles Standard and Dale Davis, authors, *Running the Factory***

> Make sure your best firefighters aren't your best arsonists!
>
> **Mark Thackeray, senior vice president, North American Operations, General Cable Corporation**

19

The Quality Manager's Role

For the plant manager who is leading the revolution to manufacturing excellence, having the right leader for the quality function is one of the early pieces of the puzzle that must be in place. Ideally the leader of the quality function should be a Black Belt in Lean and Six Sigma tools as well as a strong voice around the staff table as the customer quality advocate for the shop floor.

The quality manager's key role on behalf of the customer is to ensure the integrity of the product being manufactured and shipped and that there are zero quality escapes from the plant. Nothing is more damaging to a customer relationship than for them to have to inspect the incoming product and sort out the failures of the manufacturer. The plant's objective collectively is to prevent any quality problems from reaching the customer. While the process may well fall short of Six Sigma levels of control to the customer's specification, it is imperative that the plant team insists on internalizing the quality problems—making them highly visible—so they can be solved for the benefit of the plant's customers as well as the shareholders of the business. Now let's elaborate a bit on the significant expectations for this function.

TECHNICAL EXPERT ON QUALITY IMPROVEMENT TOOLS

Along with process engineering, the quality manager and any quality engineers on staff need to be experts on the use of the continuous improvement tool sets, which in my world means Lean tools and Six Sigma tools. At General Cable, I am proud to say that we were one of the first companies in North America to combine the toolsets for the purpose of training and certifying our own cadre of Lean Six Sigma Black Belts and Green Belts.

Further, the quality manager plays a key leadership role here in forcing the issue with the use of formal problem-solving techniques to break the long-standing paradigm of inspection. As opposed to the historical practice of quality inspectors on the shop floor or in the lab trying to catch manufacturing doing something wrong, today's quality function must be focused on what the process capability of the current process is versus a process that has been poka-yoked. It's making the critical shift in our thinking from detection to prevention. The objective is to eliminate the need for inspection by guaranteeing that the yield of the process meets all customer specification requirements. The focus is on the Critical to Quality elements of the customer's specification and the Critical to Process elements that, when controlled within spec limits, yield a quality product. Products made with this kind of discipline can be delivered right to customers' lines or right into stock in their warehouse without the cost and delay of having to do receiving inspection. Talk about a competitive advantage. Do your marketing folks think they can get an extra percentage or two of price with this kind of reliability and cost savings for the customer's business? You bet they can.

Another set of technical skills that must reside in the quality arena is ISO 9000, TL9000, TS16949, QS14000, or whatever standards exist for your particular industry. It is well beyond the scope of this book to deal with the specifics, but if certification to one or more of these industry standards is important to your business, make sure you have an expert on your quality staff who stays on top of all the requirements.

LEAD THE CHANGE FROM FIREFIGHTING TO FIRE PREVENTION

Quality professionals who have been around for a while in a traditional plant know all too well that the best firefighters in the plant are the manufacturing supervisors, mechanics, process engineers, expeditors, and machine operators. These extraordinary firefighters have given high-energy effort in the past to solve problems that were causing customer issues, high scrap, poor machine operation, or whatever. Their hearts have certainly been in the right place. They consistently took actions that they thought were in the best interest of the business. Unfortunately, in most cases they were using their experience over years of trial and error and often were simply treating the symptoms instead of the disease. They weren't really getting to the root cause. Let me illustrate with an example.

There was a particular cabling machine that required replacement of sheaves (pulleys) twice as often as other like machines because of excessive wear. Through experience, the operators, mechanics, supervisors, engineers responsible for that machine had learned over time that when the pulleys weren't replaced at least every month that the insulated conductors passing over the pulleys would be rejected. Because of the thin walls of the insulating compounds, this often resulted in "voids" in the insulating material; that is, bare copper was exposed, causing an electrical short in the conductor. Over time this team of cabling experts had installed electrical fault detectors to pick up such failures to minimize the risk that this kind of a serious quality defect would be passed on to subsequent value-added operations. Sometimes they were even able to splice or patch the cable within the requirements of Underwriters Laboratories (UL) standards—the holy grail of the industry.

I have no idea how many years this process persisted before a newly certified Black Belt quality engineer refused to believe that's the way this machine had to be operated and maintained. One of his first lessons during his training to become a Black Belt was to let the data speak unimpeded. Why? Because early in his training he learned that his early reaction to a problem often was that he already knew what to do to get the answer. Under the glare of a Master Black Belt trainer, this engineer quickly learned that his gut reaction was often wrong. Again, "In God we trust. All others bring data!"

So in his quest to find the root cause, the quality engineer did wear measurements and studied the similarities and differences between this particular machine and look-alike machines elsewhere in the plant. Not to belabor the point, but the root cause of the excessive sheave wear was that the payoff of single conductors going into the cabler was not in the exact alignment it should have been. The payoff was actually at a slight angle, which increased the friction on the sheaves that guided the wires. It was a subtle situation. The quick fix was to use guide rollers to eliminate the scrap. The correction of the root cause was to properly align the payoff. Once the payoff was repositioned into the proper alignment and bolted to the floor the cabler was put on the same preventive maintenance (PM) schedule for sheave replacements as the other machines. With the help of the quality engineer, this experience taught the operator, supervisor, mechanic, and process engineer the level of detail and the use of formal tools that are necessary to change from firefighting to fire prevention.

A small grass fire started that day in that multiconductor cell in terms of having some early believers in the new thinking and the new approach that was going to be taken on the shop floor. (These little grass fires are how you start a culture change on the shop floor!) But someone, in this case a quality engineer, had to be assertive and demonstrate a better way to address problems. The result was less machine downtime for maintenance, elimination of electrical shorts due to this cause, elimination of the potential for quality escapes to customers, and elimination of rework labor to splice, patch, or cut out the fault.

THE CUSTOMER ADVOCATE ON PRODUCT QUALITY

Sadly, in a traditional manufacturing plant the go–no-go decisions taken on marginal products sometimes are made through the eyes of the plant rather than through the eyes of the customer. For example, the quality organization may have put a hold on shipping a particular customer order because of a defect found at the packaging station. The quality manager, production control office (PCO) manager, and manufacturing manager probably have been praying over the material* every morning for a week or so until the customer finally calls and wonders why his order is late and when he'll get it.

First of all, putting off the decision helps nobody. What's called for is a collaborative effort with input from product engineering, sales management, and the customer. Delays put the customer's plans for the product at risk. It also delays manufacturing's response if the product must be replaced. Nobody wins with delays. I am aware of situations where a strong manufacturing manager basically intimidated a less than assertive quality manager to ship the marginal product to the customer without communicating with anyone outside the plant. On the other hand, I've

* Use of a material review board (MRB) is recommended with the quality manager chairing with collaborating functions participating. Typical guidelines are to disposition the suspect material within 24 hours or, in the case where a customer must be contacted, as soon as possible based on when the customer's response is received.

observed strong quality managers who took control of the situation, stood their ground against pressure from less enlightened managers, collaborated with sales and the customer, and ultimately made the correct call on the material. Obviously, it's the latter behavior from a quality manager that a plant manager is looking for.

Each industry has its own examples, but in wire and cable the issue normally comes down to physical characteristics versus the spec such as cable diameter or jacket wall thickness, electrical test results, color, and minimum requirements for the one-piece length. Let's take the case of a length nonconformance.

Perhaps due to some manufacturing issues, the cable has had to be cut such that the order came up 20,000 feet short of the total order quantity in the proper lengths. In the wire and cable industry, customers and the plant both like the longest lengths possible. It means better price per 1,000 feet for the customer and longer runs for the plant versus processing shorter lengths.

In one scenario the customer may have ordered lengths of 1,000 feet, +10 feet, –0 feet, because he had point-to-point connections to be made that were 1,000 feet apart. Anything less than 1,000 feet would be scrap for the customer. In a second scenario the customer may have bought 1,000-foot reels to minimize material handling and cost for his assembly operation. However, from a practical standpoint, for a slight price concession he could easily take 20,000 feet of cable that was made up of multiple pieces on 20 reels of 1,000 feet each. He could do this because his operation was cutting the cable into 15-foot pieces for their assemblies. At this point the sales manager negotiates a price concession, and the plant gets authorization to ship the order complete. If you were the customer wouldn't you rather be a part of this discussion as opposed to having a surprise show up on your dock late if at all? Of course. We all would. As a follow-up, the kind of quality manager we're looking for would help find the root cause of the length issue and monitor the corrective action process. He would also be sure that the customer and salespeople were aware of the results.

The quality manager is the conscience of the plant on product quality. Armed with the full support of the plant manager he must not succumb to internal pressures. He is highly visible in the plant and is the steady but relentless voice of improvement. He's making sure that Black Belts, Green Belts, and anyone else leading an improvement project are all following a formal problem-solving process such as define, measure, analyze, improve, and control (DMAIC). He educates plant personnel to create a universal understanding of what is meant by a Six Sigma level of quality. He is objective to a fault. He blows the whistle wherever he finds resistance to improvement, whether passive or active. He is the plant manager's eyes and ears in the plant on who is part of the manufacturing revolution and who is not.

> If you can't describe what you are doing as a process, you don't know what you're doing.
>
> **W. Edwards Deming, world-renowned quality guru**

> Doing things right the first time adds nothing to the cost of your product or service. Doing things wrong is what costs money.
>
> **Philip B. Crosby, author, *Quality Is Free***

20

The Human Resource Manager's Role

Of all of the great staff relationships I've enjoyed over the years, few were as special as the ones I've had with my human resources (HR) managers. They have always been my confidant and partner on all of the sticky issues that rear their heads when you have hundreds of people in the mix. The HR leader understands that each plant tends to be a microcosm of the community in which it operates, and he must be prepared to deal with any and all issues that may arise. The best I've worked with have had a calm and patient demeanor, were great listeners, and had a reputation for consistency and fairness. They also were good evaluators of talent and understood that all supervisory jobs are people jobs and require good soft skills. Individual contributors who don't understand that need not apply. While I've often worked with great corporate HR folks over the years, this chapter focuses on the role from a plant manager's perspective.

THE HOURLY ASSOCIATE'S VOICE

I always expect the HR manager to be the voice of the hourly workforce at the plant manager's staff table whether we're in a union or a nonunion plant. The salaried team, just like the indirect hourly group, is there to serve the direct hourly workforce—the only people in the plant who actually build the product and add the value that customers are willing to pay for.

For the HR managers out there who are working in a union plant and may be raising their eyebrows, let me hasten to add that, in a union plant, there are elected union officials who deserve our respect and who would be first in the loop of communicating a message that applies to large numbers of the workforce. That said, the people working in the plant are our employees, and it's our responsibility to treat them fairly, to listen with big ears, and to proactively communicate to them directly on important issues that affect their lives. My experience with union leadership is that this isn't a problem as long as they know we respect them and their roles as elected leaders. But they deserve and expect a courtesy call in advance. That's fair—and they often become our helpers across the plant in making sure the workforce at large understands the message we've sent and why it's important to them. The HR manager has to be on the point to manage this effectively.

In a nonunion plant, I also expect the HR manager to be visible and approachable. He should have a handful of loyal hourly people on each shift that he can use to dipstick the mood of the shop floor. He can bounce ideas off them, off the record, knowing that he can trust their silence until a decision has been made and announced—for example, a policy change that is being contemplated that would change the way overtime is scheduled in the plant. This skip-level communication to get unfiltered input from the shop floor is a valuable tool, and I encourage its use.

In fact, even after I got involved in multiplant operations I still liked to get direct input from the shop floor without the plant manager's filter. I think it's important at any level but absolutely essential for the plant manager. Just one caution: Don't use any information you receive to undermine the formal organization structure. Associates need to know that their identity will be protected on sensitive topics so there is no fear of repercussions. However, the content of their input is generally not confidential. If it's an issue that needs addressing, they must understand that you will put the issue into the hands of the plant manager and the HR manager to sort through it and resolve the issue. Wherever possible I always tried to get the associate to agree to a loop-closing answer from the plant manager. This helps to reinforce that I can't solve anything for them that the plant manager couldn't have solved for them had they asked. That said, there were several occasions over the years where the associate told me they'd already tried to get help from the plant manager and he either hadn't solved their problem or had never gotten back to them to close the loop and explain the answer to them—even if they would have disagreed with the answer.

The nature of the beast is that everyone's problem won't always be solved to their satisfaction. But there is absolutely no excuse for the associate not getting a comprehensive and respectful answer so that they understand why the answer is what it is. The HR manager must be the conscience of the plant on these kinds of communications and help the plant manager to reinforce his expectations on this important topic.

I grew up at Belden Wire & Cable in a nonunion environment. However, because of the hundreds of people employed by Belden and the amount of union dues that could be collected there, a union organizing campaign was never far out of our consciousness. I've been through several union campaigns during my career, and one common element led to them: Communication and respect between the shop floor and the management of the plant had broken down somewhere along the way. It might have been several generations of management before or the current regime, but the issues were generally the same. It was just a question of when the workforce became motivated enough to change their lot in life at the factory.

While I know some companies out there proactively invite unions into their operations after a long history with them (e.g., much of the auto industry), my feeling has always been that if plant personnel wanted to seek union representation, it was because of a gross collective failure of the management team. I've always thought that a plant that has a union deserves to have one.

I would always prefer to be nonunion so that there is no middleman involved in the communications between company leadership and its associates. Even still, I have rarely had difficulty working with the elected officials at the plant, regional, and national level. Once the union is already in the plant, then I've always taken the position that management will continue to run the business with the union leadership as our partners. Together we will still commit to taking care of our customers and shareholders and educate and train the union leadership. Further, we will share the key metrics of the business so that they really understand how the plant is doing and where the targets are to get better. Union leaders know that the enterprise must remain viable to continue to provide jobs and to continue the cash flow from union dues. The international unions are in fact very large businesses in their own right with their own management structure and salary structure; it also is big business. Ironic, right?

I've spent a lot of time on this first topic for the HR manager. Is there any question that labor relations are job number one?

SERVICE-ORIENTED STAFF OF EXPERTS

The HR manager and his staff must be functional experts on a variety of subjects. This is especially important in HR because seldom are there others around the staff table who have ever worked in HR. I have no axe to grind on whether this is a good thing or a bad thing. It's just the way it has been throughout my career. It is so much better to have a few experts on topics such as employee benefits, compensation, labor law issues, Equal Employment Opportunity Commission (EEOC) rules, and Health Insurance Portability and Accountability Act (HIPPA) rules. Can you even imagine having to educate and train all of the supervisors and managers in the plant so that they could field any question on these topics and more? Talk about chaos and confusion.

Of course, it is absolutely essential that the HR manager makes certain that he, in fact, does have a team of experts. Using suppliers of the products being used to do the training (e.g., the dental plan provider) is always a good idea. Over time, however, sufficient cross-trained experts may be available in-house to do the training. Annual updates as programs are modified are essential as well. In any case, it is the HR manager's job to make sure that his staff is always fully competent with the current state of affairs in their respective areas of HR. Also, since most plants aren't large enough to have all of this expertise on staff, the HR manager typically has to use division or corporate experts as an extension of his team.

The important thing for all supervisors and managers to know is the person's name and extension to call for specific HR questions. And, of course, the HR representative must then be completely trustworthy in responding with a timely answer to the associate's question. This kind of behavior on their part helps to reinforce the integrity and dependability of the entire management team.

Several times in my career I witnessed that the HR team did a great job if the workforce had a need between 8:00 a.m. and noon and again between 1:00 p.m. and 5:00 p.m. This typically was more than adequate for the first two shifts of a three-shift operation, but it did nothing for the poor souls who earned their living on the third shift and typically started working late at night and finished up anywhere from 6 to 7:00 a.m. Would you feel like a second-class citizen if you had to wait around for an hour or more after working a full shift to see someone in the HR department? You bet.

The kind of HR manager we're looking for is the one who immediately makes certain that all associates have ample access to his team of experts. And if there's an exception to that, all one has to do is to let him know that, and an expert will arrive at the associate's workstation at whatever time has been agreed.

PLANT MANAGER'S CONFIDANT

The HR manager, by virtue of his position, must fill the role as the plant manager's confidant. I always encourage that this relationship be very broad based and to even include issues that may be outside of the normal realm of HR. The HR manager must be beyond reproach when it comes to any and all confidential matters such as wage and salary administration, employee discipline, peer group performance reviews, development plans, succession planning, and lawsuits against the company. If there is anything less than 100% credibility in the plant with regard to the HR manager's integrity, ethics, or morals, then the first thing the plant manager should do is to find a new HR manager.

It's also important that the HR manager be straightforward with his peer group that his role as the plant manager's confidant is a common one he is expected to play. That said, he will promise the sanctity of any communication between him and any other member of the plant staff with one notable exception: when withholding information from the plant manager would not be in the best interest of the business (i.e., its customers, shareholders, or employees). In those cases the associate must understand that this responsibility trumps any reason he may have for keeping a secret.

The plant manager will work with his HR manager much like a peer on plant processes such as salaried performance reviews, associate development plans, succession planning, salary planning, and discipline. That's why that when the door is closed the HR manager must be allowed to say whatever is on his mind.

Take the case of the annual performance reviews. The HR manager becomes the sounding board for the plant manager to bounce off his thoughts on each of his direct reports. The plant manager should seek out the division- or corporate-level HR executive for the same kind of input for the plant HR manager's review. This should include a discussion of results versus objectives but also the soft side of management. For example, does the manager in question play well with others? Is he a good teammate when the plant manager isn't around, or does he try to be controlling with his peer group? How do his direct reports feel about his leadership? Is there any worrisome strain between him and any of his peers? If his name was in a promotional announcement today how would his employees react? For instance, "Boy, am I glad to get rid of him," or "That's great news; it couldn't happen to a more deserving person."

When the discussion moves to succession planning, the plant manager may be thinking about cross-training his high-potential materials manager into a manufacturing role to round his experience in preparation for someday managing his own plant. What's the HR manager's take on that? Does he have the makeup necessary to progress? Does he have his people's respect in his current role? Are there specific technical training or behavior development "flat spots"? How will we address these issues in his development plan for the next year to assist in getting him ready to assume greater responsibility? His salary is a bit low to his peer group considering his performance level and his potential. HR manager, please help me to put a salary plan together that, over the next 18 months, improves his position to X% of the midpoint of his range and corrects the slight inequity with his peer group.

These are conversations the plant manager simply cannot have with anyone other than the HR manager in the plant. That's why the HR manager must be able to perform this part of his role as capably as he does an interview with an hourly associate.

POLICY ADMINISTRATION

Most plant policies come under the purview of the HR manager. He must make sure that the management team has a good understanding of plant policies as well as their intent. In other words, the other members of the plant staff should defer to the HR manager if they are confronted with a question they aren't confident to answer. Often the staff member can simply call for clarification and provide the answer immediately to the associate. In other cases it may be better to ask the HR person to provide the answer directly to be sure the communication is accurate and clearly understood. If possible the staff member should be present when the answer is delivered by HR. For example, what is the policy with regard to seniority when the associate has had a break in continuous service?

When we think about policies in the plant we often think about employee discipline. Most plants have what is called a progressive discipline policy that administers discipline based on either the frequency of occurrence, the severity of the incident, or both. For example, two absences that are called in for the same month might be cause for the employee's supervisor to have a conference with the employee to determine what's going on that caused the absences. The results of that conference might be a review of what the policy is if that behavior were to continue as well as a verbal warning. Along with the verbal warning the supervisor should also be sure the associate understands that even though they called in 1 hour before the shift started, per policy, that it still caused the supervisor to find a replacement to cover his job on that shift and that the company had to pay time-and-a-half to get it done. Typically the next violation would be a written warning that goes into the associate's personnel file. A third infraction might result in a 3-day suspension. A fourth violation might result in termination as an undependable, unemployable person in this plant.

In another situation, two associates might get into a fight inside the plant. The policy for this could be termination on the first offense once an investigation is done and witnesses, including the two involved in the fracas, are interviewed to determine who instigated and who escalated the altercation. The outcome will likely be termination of both parties unless there is compelling testimony from witnesses that one of the two combatants was clearly the aggressor and that the victim had simply been trying to protect himself from injury.

The point is the kind of infraction is a key factor in determining whether the punishment is immediate or progressive. The HR manager leads the objective, fact-finding investigation and is nonjudgmental until all the facts are in. In the case of a suspension or a termination, the HR manager always makes sure that the plant manager and the appropriate staff level manager know and agree with the HR manager's intended action and that the employee's direct supervisor is included in the interview where the discipline is administered.

The worst thing that can happen next is for an associate to come forward and claim that another employee did exactly the same thing as the one who was just terminated but got only a 7-day suspension. These things simply cannot happen. That's the critical reason HR has to be involved in any discipline. They are the only people in the plant who have a view of the whole field and can ensure consistency in the administration of discipline across the various departments or cells in the plant. The plant manager counts on being able to talk about consistency on this topic with anyone in the plant.

ENVIRONMENTAL HEALTH AND SAFETY

In a large plant this function might have a staff-level manager heading it up. However, in most plants this position will be filled by a technically competent engineer who is knowledgeable with regard to environmental health and safety regulations. For example, he would be the resident expert on any permits required in the operation to meet U.S. Environmental Protection Agency (EPA) requirements. The same would be true on Occupational Safety and Health Administration (OSHA) safety issues such as the use of proper protective equipment like safety glasses or ear plugs.

Over the years I've seen the environmental health and safety function as part of plant engineering and maintenance or perhaps risk management at a corporate level. But most often I've seen it lined up as a part of the plant HR department. That's certainly my preference. Why? Because safety is a people issue so I want my employee advocate leading this charge but with the support of a qualified engineer who provides programmatic, technical solutions where required. Safety also is the cornerstone of the culture change that we seek, and I want my culture change zealot leading the way.

In the case of environmental, the environmental health and safety person is technically competent to deliver cost-effective compliance to or prevention of environmental issues whether in regard to water, air, or ground contaminations issues. And the HR manager's and plant manager's vocal support are important to give the workforce the highest level of confidence that their leadership is conducting its business with the legal, ethical, and moral values they would respect and admire.

Also, as the plant manager, I'm going to have a close personal relationship with the environmental health and safety person on environmental matters. To show this expectation, the level of interest, and commitment the conversation might go something like this: "On my watch there will be no breeches of the law and nothing getting kicked under the rug for another day. I want to know what all of the issues are and understand the priority set for action where required. I want you to know that you have my complete support and will always have an open place on my calendar if you need my help. I'd like you to prepare an inventory for me of all of the things on your agenda. One of those I hope will be a discussion about how we can prevent future issues by monitoring the front of the pipe for proactive changes as opposed to continuing to monitor the end of the pipe where we have to deal with remediation. Where there are issues that we need to deal with, I'd like you to prioritize them. For example, a #1 item must be corrected as soon as possible so I'll need to know the cost of #1 item so I can plan that into my forecast for expenses or capital spending. A #2 item means that you either need to collect more technical data or perhaps consult with the EPA for suggestions or to plan a longer-term corrective action so that proper planning can be done. A #3 item goes on the watch list for monitoring to determine any future action that may be required."

You can see the importance that must be placed on environmental health and safety. In the case of serious environmental issues, it is one of the few things under a plant manager's purview that if not done properly might put him in jail. You can expect that he'll take it very seriously.

CULTURE CHANGE ZEALOT

The final key role I'll mention for the HR manager is that he is the culture change zealot of the staff. Suffice it to say that the HR manager must be joined at the hip with the plant manager and share the vision of the culture that must be created throughout the plant in order to achieve and then sustain manufacturing excellence. Certainly it starts with his own challenge to get his own staff in alignment, but it goes well beyond that. He will become a critical partner to the plant manager as the training and communications needs become apparent once the torch has been lit to start the revolution, first on the shop floor and later in all of the offices.

As you saw in Chapters 11, 12, and 13, HR owns manufacturing principles 10, 11, and 12 on communications, training, and culture change, respectively (i.e., Operator-Led Process Control [OLPC]). Without strong and persistent leadership coming from the HR manager the plant will simply never be able to sustain excellence because the culture will not be in place to do so.

> Failure to establish a sense of urgency is the single largest mistake people make when trying to change an organization.
>
> **John Kotter, author, *Leading Change***

21

The Finance Manager's Role

The finance manager in many companies is the one person around the plant manager's staff table who often doesn't report directly to the plant manager but rather is a dotted line report. This is very simply a control mechanism to eliminate the prospect of a plant manager pressuring the finance manager to do something improper with the reporting of the numbers or worse to engage in fraudulent practices under duress from his direct supervisor. Instead the finance manager has a hard line report up through the finance structure to the chief financial officer (CFO).

In addition to this control issue, a second major reason for this structure is to ensure that the finance manager has a technically competent supervisor. Similar to HR, the finance function has its own set of rules, e.g., Generally Accepted Accounting Principles (GAAP), and laws or regulations, such as Securities and Exchange Commission (SEC) requirements for public companies. Most often the plant manager has not come up through the finance ranks, nor is he a certified public accountant or a degreed cost accountant. As a plant manager and as a corporate leader I always respected this structure because it helped to ensure that the operation was in complete compliance with all of the proper accounting requirements. Of course, since Sarbanes-Oxley legislation was passed in 2002, the plant manager and the finance manager have had to sign-off on the quarterly financial reports verifying that they are in compliance with all of the proper accounting and reporting requirements. It's fair to say that line executives are taking an interest in understanding more of the details behind the numbers since this legislation was put into effect.

Now that we've clarified the typical structure for the finance arm of the organization let's focus on how the finance manager also participates as a member of the plant manager's staff.

ESTABLISHING RAPPORT

It's important in the early days of the relationship for the finance manager and the plant manager to have a lengthy discussion to establish a rapport based on mutual understanding and respect for each other's job requirements. Because of the different reporting structure it is important for the finance manager to hear the plant manager's assurances that they are on exactly the same page with the CFO. For example, the plant manager's first discussion with the finance manager might go something like this:

Plant manager: I am glad to have you on my team to be sure we keep everyone and everything on the rails with regard to proper accounting and reporting. I just want you to know that you have my full support to hold the rest of the staff's feet to the fire to be certain that nobody in this plant is ever guilty of anything illegal or unethical when it comes to our accounting for and reporting of the numbers. If you ever suspect any questionable practices by any of our people I expect you to inform me immediately so that together we can take the appropriate actions. I also respect the fact that in any situation like this you have an obligation to report the same thing to your direct supervisor immediately. All I ask is that you review it with me first so that I don't hear it for the first time from my boss, who will likely be the first phone call your boss makes just as soon as he hangs up the phone. Deal?

Finance manager: Absolutely. From my side I would simply ask for the same courtesy if you have any kind of an issue with me that you would discuss it with me first before discussing it at the next level. OK?

Plant manager: You bet. I'm looking forward to working with you. You'll be a great part of our team here. I'd like to move on now if you don't mind and discuss what my expectations are relative to how you participate with your peer group around the staff table.

THE FINANCE MANAGER'S ROLE WITH THE OTHER PLANT STAFF MANAGERS

I expect the finance manager to be just as reciprocal with the plant manager's direct reports. For example, if the finance manager sees inventories growing without explanation he should raise the issue directly with the materials manager before taking the issue to the plant manager. The same thing would apply with the manufacturing manager if the finance manager develops a concern about manufacturing not managing variable expenses effectively. It's just common courtesy and is important to demonstrate that he is part of the team regardless of formal reporting lines.

That said, if the finance manager becomes aware that the manufacturing organization is not reporting all scrap on a timely basis or is not processing engineering change orders in a timely way regarding current cost accuracy or other matters that are considered financially material, then by all means the plant manager would expect to be pulled in sooner rather than later. A heads-up to the appropriate staff-level manager is fine as long as there is a very short window for that manager to either solve the problem or to come clean with the plant manager. When it has to do with the integrity of the numbers the finance manager must play his role without reservation. Again, that's why he reports up through the CFO to the audit committee of the board of directors.

KEEPER OF THE METRICS

Often the plant manager will look to the finance manager to collect and report important metrics. Where it pertains directly to the operating statements, that's a no-brainer. However, I always asked my finance counterpart to oversee nearly all of the plant measurements to guarantee consistency in the use of the proper process or formulas in the calculations—for example, creating the standard work for the calculation of the overall equipment effectiveness (OEE). This assures not only consistency but also that the fox isn't watching the hen house, or cooking the books on a particular metric.

Also, since the finance manager collects and issues reports on a daily, weekly, and monthly basis, why not add on the handful of other important plant metrics in the interest of standard formats, timely distribution, and so on? The finance manager also becomes the unofficial mentor to other members of staff to be sure that all understand the purpose of each metric and what they're looking at in terms of performance.

CAPITAL INVESTMENT WATCHDOG

The finance manager is the financial conscience of the organization. In addition to normal accounting functions, I always relied heavily on my finance person to carefully guide the process of compiling the financial support for capital investments. Further, he had full latitude in challenging the justification for such investments to help make certain that the project sponsor's return on investment forecast was doable and that the improvements promised were trackable to the income statement.

Follow-up or post-audits are also an important part of the process to check on the actual deliverables versus what was promised in the capital expenditure review (CER) process. The finance manager also is a good watchdog with the sourcing organization to be sure that the company is receiving the benefits of their corporate purchasing power and that an appropriate level of competition (typically at least three bids) was solicited in the bidding process. Until the finance manager had signed off on the integrity of the numbers being presented I would not review or sign the proposal. That's how much I depended on and valued the input.

SYSTEMS INTEGRITY ZEALOT

As we've already discussed in previous chapters, the finance manager must have his oar deeply in the water on systems integrity. If material usage, scrap, inventory, and any other transactions are not reported properly in the authorized formal system then he has no chance of having accurate costs in the system and accurate financial reports going up or down the structure of organization. As such he has a vested interest in auditing and being sure that he isn't part of a garbage-in, garbage-out process. His career might depend on it.

SHOP FLOOR VISIBILITY

My best finance managers over the years were those who were not chained to their desks. I love to see the finance manager visiting the shop floor. What a great skip-level contact to check the understanding and integrity of shop floor reporting. For example, he might go up to a machine operator after looking at the nearby trash bin and ask, "Hi, Tony. Good to see you. Hey, I've got a question for you. When you change tape pads on your machine, what do you do with the leftovers?"

Tony might reply, "Hey, Keith. Must be slow up in the office today, huh? Anyway, it's good to see you, too. As for the leftover tapes, all operators are trained that tape leftovers are returned to the raw stores inventory unless there is less than ½ inch left on the core. In that case, they are thrown in the trash. I'm told that the cost of the 'leftovers' is accounted for with a usage factor in the bill of material."

Some of you are thinking, "Yeah, right." But is this not a great way to get a good audit of the operator's understanding of an important material usage issue? And isn't this the answer that you want to get if you ever hope to work in a world-class operation? The finance manager can also get a feel for whether he's just found the only operator who understands and complies or if there is broad-based understanding.

Unfortunately, the finance manager may also get a response such as this from another operator, Ray:

Ray: As for your question about leftover tape, it kind of varies by operator. Some are really finicky about running the tape down as far as they can before throwing away the rest. Others don't like the time it takes to change the tape pads as they run out one at a time so they aren't that particular about it.

Keith: What do you mean?

Ray: Well, a lot of operators run until the first pad runs out, and then they change out all of the other pads to avoid more interruptions later.

Keith: Wow that sounds expensive! Do you mean they do this whether they're running a two-tape job or a four-tape job?

Ray: Yep, that's usually how it goes.

Of course, this audit could just as easily be done by the first-line supervisor, the sourcing person, the manufacturing manager—anyone really. But it has great impact when the senior finance manager at the facility shows this kind of interest and engages directly. His next stop, of course, is to the supervisor's office to report on the conversation he just had with the two operators.

The second thing the finance manager might do would be to drop the manufacturing manager and materials manager an email reporting his findings and that he had already discussed it with the appropriate supervisor. He might then suggest that a review be made of the training plan to be sure the standard work is correct and that every operator has signed off on the correct procedure to use. Then it's up to the supervisors to audit with sufficient frequency to ensure compliance while knowing full well that the finance manager will be back at some unknown time in the future to check the trash bin again and ask some questions.

The finance manager may also do an impromptu audit on scrap reporting, scales calibration on an important raw material scale in the compound mixing room, or the timing of the entry of raw material receipts into the formal system. In short, there are lots of ways that having the finance manager's eyes in the shop can be helpful to reinforce the value of company assets, tight financial controls, and individual operator responsibility to manage their materials like they had to buy them with their own money. Clearly, this is not what is often referred to as the "bean-counter" role in the factory that, unfortunately, some still play.

These important roles for the finance manager are just some examples of how proactive and supportive this key team member can be if he understands the plant manager's expectations and is cut from the right cloth to work in a factory. Those who prefer to live in an ivory tower need not apply.

> The greatest danger for most of us is not that our aim is too high and we miss it but that it is too low and we reach it.
>
> **Michelangelo (1475–1564)**

22

Sustaining Manufacturing Excellence

I hope by this time that you have a full appreciation of the focus, the passion, the disciplined processes, and the strong and steadfast leadership that are required to achieve manufacturing excellence. It certainly is not for the half-hearted, and it is not for someone looking for instant accomplishment that can be checked off this month's to-do list.

I also hope it is crystal clear that seeking excellence is all for naught unless the infrastructure is put into place with all the controls necessary to sustain it—ala the 12 manufacturing principles—and unless the present and future leaders of the business are fully committed long-term. When I say committed, I mean not just to episodically using Lean tools to churn out a short-term improvement (the easy part) but also to change the culture of the business (the hard part). These are the two most critical factors in my view to be able to sustain for a long enough period of time to effect real culture change and long-term operating results.

Constancy of purpose and focus is essential from the leadership at the corporate office on down to the shop floor. For example, a terribly broken Stage 1 plant may take 10 years to progress into Stage 3 and Stage 4 kinds of performance and behaviors. These plants simply don't have the infrastructure and discipline in place to skip steps in the rebuilding process. They are Stage 1 for multiple reasons. It takes years of hard work to put the necessary foundation in place. It makes no sense to start building the roof until the foundation and the walls are solidly up. Taking shortcuts won't work. On the other hand, a good performing, traditional Stage 2 plant may move through Stage 3 and into Stage 4 in just a few years. This progression, of course, depends on lots of factors. And it's also why Operator-Led Process Control (OLPC)—the culture change—is so critically important.

Changing culture is a race. Leadership must win the hearts and minds of the critical mass of the hourly associates through persistence and steadfastness of purpose. There will be missteps along the way that will test everyone's resolve. For example, how will plant leadership react when there is an interruption in the newly designed flow in a sold-out cell and the phone is ringing off the hook from the customer service department due to late orders? Will they throw out the value stream map and bring in the fire hose? Will they patiently, but with urgency, determine the root cause of the interruption and make the necessary correction, or will they instinctively return to batch manufacturing mentality and load up on inventory just in case it happens again? This is just one example of the many challenges that will be faced in the early days of the manufacturing revolution that will test the will of the leadership. And you can be sure that the hourly associates will be watching to see how management reacts.

Unfortunately there is often a relatively small window of time to accomplish sufficient change to create positive momentum. Management changes are a way of life. People move on to other opportunities inside the company. Some leave for greener pastures in a new company. There tends to be constant churn at the management level. Unfortunately, these frequent changes often derail continuous improvement agendas.

That's why showing some early results with both performance and examples of OLPC success is important to gaining credibility for the manufacturing excellence initiative. I encourage you to select an area of low-hanging fruit that also happens to have a nucleus of open-minded hourly associates who want to try the new way. Starting out with a complex problem most likely will result in a false start and a lost opportunity. Be purposeful with the selection of your early improvement projects. Once you have had and celebrated a successful pilot project, expand use of "the new way" methodically around the plant one cell at a time. That's how you start a revolution. Once the broad base of hourly associates has begun to change how they think and how they work there will be a powerful inertia not to resort to old habits even after subsequent changes at the management level. Culture change is like a spreading grass fire that breaks out in enough areas of the plant until there is finally critical mass. At that point there is no stopping a work force that is on fire. Creating this change on the shop floor is critical to being able to sustain manufacturing excellence.

As we said earlier, management changes are a way of life in corporate America. On the other hand, the hourly team usually is much more stable. They typically seek to make a good living wage while staying in the same town to raise their kids and be close to the grandparents. In short, they are less mobile. As a result, they have a higher vested interest in the success of the local plant. Ideally they'd like the plant to be successful and grow jobs for their friends and neighbors. They'd like to feel secure that they will be able to retire from the plant one day and continue to live comfortably. They may even dream of their children or grandchildren having the opportunity to work there someday and stay in the area. Changing the culture will cause them to take real ownership of their work and help them to realize their dream of long-term job security for themselves, their family, and their community.

As you know, each company and each plant has its own history and its own culture. As my Association for Manufacturing Excellence (AME) friend Joe Barto likes to say, "When culture and change collide, culture always wins." This is why most continuous improvement initiatives fall flat after a couple of years. Leadership gets weak in the knees because of the whining noise that emanates from down the organization or because something else comes up and management's focus gets shifted to something else. This phenomenon is otherwise known as flavor of the month and why hourly folks are always skeptical when a new leader swoops in and says, "Let's do Lean. From the associates' point of view, "The last guy said let's do something else. Now he's gone and we didn't get the last thing done either. This too shall pass."

To break this paralysis, starting such a forever initiative as continuous improvement cannot be up for a vote. To repeat the story from Chapter 1, once the strategy has been agreed to by the corporate leadership team and the board of directors, then everyone who wants to stay in the company has to be committed. The voting is over. When I launched our manufacturing excellence strategy at General Cable in 1999 I called a special conference with all 28 plant managers and the corporate operations staff together for a major presentation and setting of the new direction and the higher expectations.

I could see in the eyes of my group three different messages. About a third of the group had a sparkle in their eyes that told me they were already onboard with me and ready to rock and roll. Not surprisingly, these were my best people already. The rest of the group was split about evenly. There was a significant group of people—again, about a third—with questions in their eyes. I took this as a positive. They were silently telling me, "I'm interested, but I don't yet understand. I will need lots of training because I don't know how to do what you are describing. But I will try to learn."

The last third of the group had eyes that suggested, "I can outlast you. I've been with this company a lot longer than you have." Others in this group just looked totally detached like they were having an out-of-body experience. I interpreted this group as communicating their lack of interest in changing. They were already convinced that their people back home wouldn't buy it. I knew this group was likely to resist, either actively or passively.

Many of you have probably read Jim Collins's book *Good to Great.** He suggests a simple but powerful way to think about your organization at any time, but especially when you launch such an important initiative as manufacturing excellence. To paraphrase one of Collins's key messages in the book, a great leader is going to surround himself with the right people who will help lead the change that is envisioned. These decisive changes should be made in the first several months. Failure to act quickly only slows progress and causes frustration. Here are the three things Collins recommends:

Make sure you have the right people on the bus.
Make sure you get the wrong people off the bus.
Make sure you have the right people in the right seats on the bus.

The last slide in my presentation at this special plant manager conference was aimed at everyone in the audience but especially at this last group of skeptics. The slide was a picture of a leader in front of his staff having just finished a presentation. The caption read, "All those not in favor, signify by saying I QUIT!!!"

That was my way of communicating that the voting was over. We were no longer going to think and work like we had in the past. Everyone will be expected not just to think and act differently themselves but to also lead their plants and their staffs and teach them to think and work differently. I told the audience that a year from now there would be a number of new faces in the audience because some would be incapable of change and others simply wouldn't want to try. Over the next 18 months

* *Good to Great: Why Some Companies Make the Leap...And Others Don't.* 2001. Harper Business.

or so about one-third of the plant managers either left on their own or were replaced.

This was the first real moment of truth for me as some of the plant managers being excused had been with the company for a long time and, in a very traditional plant sense, had been recognized for doing a good job. I was blessed to have a chief executive officer (CEO), Greg Kenny, who never blinked. He stood firmly in support of the important work we were doing. That strong support of what we started in 1999 is still going strong now more than 4 years after my retirement. I hope all of you who are leading the charge in your company are fortunate enough to have this same commitment and level of confidence as I enjoyed. It makes all the difference in the world.

This is in contrast to those of you who may be in the unfortunate position of having to start your manufacturing revolution from the bottom up. Our revolution at Belden started in 1987, and we were about half-pregnant with our major cellularization effort when there was a huge shake-up in our business. Fortunately, our parent company, Cooper Industries, had noticed that something was going on at Belden-Richmond and they wanted to know more. What was going on was that finished goods inventory was going down, inventory turns had doubled, and service levels were up 15%. Rather than the new leader coming in and changing direction, he wanted us to accelerate the process. His name was Nish Teshoian. I'll never forget what he said to me when I expressed my concerns in my first meeting with him: "Larry, your organization is showing the most improvement in the company right now. Who in their right mind would ask you to stop doing what you are doing?"

So be bold. Take on a pilot with a high probability of success and do something great. And then expand the pilot. The best part of starting from the bottom up is the early results captured from the low-hanging fruit make it possible to pay as you go and make the costs of manufacturing change transparent to the financials in the short-term. Go for it.

As you can see from Figure 22.1, the results accomplished by our manufacturing excellence initiative at Belden resulted in performance improvements that simply would not have been possible with incremental improvement goals and the use of traditional methods. Bold objectives and unbending commitment to excellence are what created step-change improvement in performance.

Journey to Manufacturing Excellence

Summary of Results — Belden Wire & Cable
Base Year - 1988

	% Improvement 1988–1992	
	% Range	**% Average**
Delivery Perfomance	8 to 91	56
Raw Material Dollars	6 to 76	51
Raw Material Turns	48 to 237	92
WIP Dollars	27 to 52	47
WIP Turns	39 to 324	108
Cycle Time	25 to 75	50
Travel Time (Material Handling)	53 to 86	60
Indirect Headcount (Material Handling)	32 to 60	47

Additional Notes:

80.3% of machines moved from process departments into manufacturing cells built on common routings (in today's terms, around value streams!).

Cells helped to eliminate two inefficient, poorly performing plants plus an additional 89,000 sq. ft. in remaining plants. Toal of 429,000 square feet reduction. No capacity was reduced as a result of these plant closures.

FIGURE 22.1
Journey to Manufacturing Excellence, BWC.

THE REST OF THE STORY

Now let's fast forward to my last 6 years at General Cable.

As the North American operations team grew in confidence and conviction, the performance in the plants had improved significantly. The 12 Principles of Manufacturing Excellence and the Manufacturing Excellence Audit had truly developed into a proven system for achieving and sustaining continuous improvement and culture change (Figure 22.2).

In earlier times there were so many basic control issues in the plants that manufacturing had to stay very internally focused. But now it seemed like the better the plants did in solving their own problems on the shop floor and in the office, the more obvious it became that many of the problems being found on the shop floor had been sent there by someone else: a corporate purchasing person, a design engineer, an accountant, a customer service person, or a salesperson. The dysfunction was running rampant, and most of the people now creating the problems

Journey to Manufacturing Excellence

Summary of Results — General Cable Corporation
North American Operations
Base Year - 1999

	% Improvement 2000–2006	
	% Range	**% Average**
Delivery Perfomance	14 to 78	38
Raw & WIP (RIP) Turns	25 to 120	77
DPMU	29 to 97	61
Manufacturing Excellence Audit	20 to 175	60

Note:

Industry Week Best Plant Winners improved MEA scores by an aveage of 66% ranging from a low of 37% to a high of 175%.

In 2000 the range of scores on the MEA was 32 to 68 with an average score of 53.

In 2006 the range of scores was 66 to 93 with an average score of 81.

3-Year Bold Objectives	**Cash Generated**
Scrap improved by 30%	$ XX,XXX,XXX*
RIP Turns improved by 65%	$ XX,XXX,XXX
Finish Goods Inventory	
Reduced by 32% (while sales were down by 8%)	$ XX,XXX,XXX
	$ 122,000,000

Total cash released in N.A. Operatons was critical to the company's survival through the first half of the decade after the events of September 11, 2001.

Service Level Improved from 86% to 95%.

Safety Improvement of 76% on the OSHA incident rate – from the worst in the industry to among the very best.

Note: Since my retirement I understand there has been another 55% improvement!!! Safety performance is now among the best in the world!

* Specific breakout of these results withheld for competitive reasons.

FIGURE 22.2
Journey to manufacturing excellence, General Cable Corporation.

were naïve to the negative effect, in some cases havoc they were causing in the plant.

But one really important thing had changed. Manufacturing performance had improved to the point that the business team and other functional leaders at the corporate level had taken note of what was going on in manufacturing and were developing newfound respect for the function and the people in it. Manufacturing could finally speak with a loud voice at the leadership team table without being defensive. Managers in other functional areas began to listen with a more open mind about the ways they were hurting company performance and took more of an interest in helping to make the business better. Here are some examples of what I mean:

- The design engineer who saw and understood the results of a process capability study on one of his product designs which was so on the edge (i.e., a Cp <1) that the plant had a very poor chance of ever producing that design profitably.
- The inside sales person who finally understood that taking an order for 2,000 feet of a special design didn't make good business sense and always became a disaster in the plant. He thought he was quoting a 40% gross margin but in reality was losing that much.
- The salesperson who began to push back with a good customer who was used to blaming his problems on manufacturing while never sending in a sample to help with problem diagnosis and was now being challenged to prove it instead of automatically getting a credit for the alleged problem. Because of our poor history we were still being taken advantage of by a few customers. So the sales organization stepped up.

Finally, manufacturing had earned a seat at the strategic planning table, that is, an invitation to participate in the discussions about what was best for the business. Manufacturing was now being viewed as a vital resource in helping to build customer relationships, and the journey for excellence was evolving into an enterprise-wide endeavor.

The efforts by the supply chain organization were also paying off now. A much more robust sales and operations planning (S&OP) process was coming together with much stronger support by the business general managers and the corporate sales leaders. The company had finally begun to operate as a Stage 3 company with some early signs of Stage 4 kind of external integration and support. It was very gratifying to know that this was in large part enabled by the plant managers and their teams, along with the corporate operations staff, working hard to radically impact the business in a very positive way. I sincerely hope that each of you who is interested enough in manufacturing excellence to read this book will one day experience this feeling of self-fulfillment that comes with forever changing the course of your company.

I'll close with what I've learned about leadership. The best leaders:

- Maintain a healthy level of dissatisfaction with the status quo every day.
- Are relentless in their quest for continuous improvement.
- Have high expectations and hold people accountable for results.
- Have a vision of excellence that is communicated broadly and frequently.
- Communicate with passion a clear priority set.
- Don't pass the buck when there is bad news.
- Run to the problems and deal with them head on.
- Can see opportunities that others can't.
- Are the ones who care the most and make everyone around them better.
- Know when it's time to leave.

Every morning in Africa, a gazelle wakes up. It knows it must run faster than the fastest lion or it will be killed. Every morning a lion wakes up. It knows it must outrun the slowest gazelle or it will starve to death. It doesn't matter whether you are a lion or a gazelle. When the sun comes up, you'd better be running.

Taken from a "Successories" picture, which hung in the author's office for many years

Section 3

Appendices

Appendix A: The 12 Principles of Manufacturing Excellence

1. **Safety is the cornerstone** of a high-performance plant. It is characterized by high associate awareness and involvement—effective associate training, ergonomically sound work environments, vigorous investigation, and root-cause elimination of unsafe acts and conditions—and results in zero lost or restricted time accidents.
2. **Good housekeeping and organization** are expected at all times. Use of the 5S (sort, set in order, standardize, shine, and sustain) technique is required: a place for everything and everything in its place.
3. **Disciplined use of authorized formal systems** is required to ensure data integrity of bill of materials (BOMS), routers, labor, scrap, and inventory records. Use of inaccurate data results in financial and customer service surprises and causes poor decision making.
4. **Preventive/predictive maintenance systems** will be routinely used to plan and schedule equipment and facility maintenance. An undependable plant delivers poor customer service and disappoints shareholders.
5. **Process capability will be measured on all key processes** with a minimum process control expectation of >4 sigma, 1.33 Cpk. Use of statistical tools leads to a reliable environment with predictable outcomes on quality, cost, and service. The ultimate objective is to achieve theoretical levels of capacity and material use with process control approaching 6 sigma, 2.0 Cpk.
6. **Operators are responsible for product quality** and will not knowingly pass defective material to the next operation. The objective is to have zero escapes of poor quality product to the next operation or to an external customer.
7. **Product will be manufactured on time** to the original, agreed-upon delivery promise. Manufacturing's delivery performance is critical to growing the business profitably. It is a measure of the reliability of the shop floor and is an essential part of providing service that ultimately delights the customer.

8. **Evidence of visual management** will be prominent throughout the plant with key metrics at each work center. Kanbans, Andon lights, audio alarms, color coding, and other visual techniques will be used as appropriate.
9. **Continuous improvement** will result in relentless productivity improvement, year over year, forever. It is critical to the long-term job security of everyone to maintain a low-cost and competitive manufacturing cost structure.
10. **A comprehensive, purposeful communications plan** will be in place and executed in every plant on a rolling 12-month basis. Better-informed associates take more interest in and make better decisions for the business.
11. **A comprehensive, purposeful training plan** will be in place and executed in every plant on a rolling 12-month basis. Fully competent associates work safely and deliver quality products on time at a competitive cost.
12. **All associates will help** to create and sustain a shop floor environment where the operator is in control of the process, known as Operator-Led Process Control (OLPC).

Appendix B: Manufacturing Excellence Audit

Disclaimer and terms of use: The Manufacturing Excellence Audit (MEA), Example of a Communications Plan, and Example of a Training Plan are Excel workbooks with standard working formulas. Their purpose is to provide the user with sample templates for reference. User agrees to release author and support personnel from any and all liability associated with the use of the formulas used in these workbooks.

Refer to Appendix B on the CD-ROM inside the back cover.

Appendix C: Manufacturing Excellence Reading List, 1986–2010*

REQUIRED READING

Orientation

*Restoring Our Competitive Edge—Competing Through Manufacturing, *Harvard Business Review*, Steven Wheelwright and Robert Hayes
*The Goal: A Process of Ongoing Improvement, Eliyahu M. Goldratt and Jeff Cox

Manufacturing

**World Class Manufacturing: The Lessons of Simplicity Applied*, Richard J. Schonberger
**Competing against Time: How Time-Based Competition Is Reshaping Global Markets*, George Stalk and Thomas M. Hout
**Reinventing the Factory*, Roy L. Harmon and Leroy D. Peterson
**Manufacturing: The Formidable Competitive Weapon*, Wickham Skinner

Lean—General

**Just-In-Time: Making It Happen: Unleashing the Power of Continuous Improvement*, William A. Sandras
**Just In Time for America*, Kenneth A. Wantuck
**Just-In-Time*, Walt Goddard

* The * means this was the original list that helped start the manufacturing revolution at Belden in the mid-1980's and has continually been updated as new titles are added.

Lean Thinking: Banish Waste and Create Wealth in Your Corporation, James P. Womack and Daniel T. Jones

A Study of the Toyota Production System from an Industrial Engineering Viewpoint, Shigeo Shingo

Fast Track to Waste-Free Manufacturing—Straight Talk from a Plant Manager, John W. Davis, Productivity Press

Lean—Tools

Learning to See, Mike Rother and John Shook

TPM for Supervisors Shopfloor Series, Japan Institute of Plant Maintenance

5S for Operators/5 Pillars of the Visual Workplace, Productivity Press Development Team

Standard Work for the Shop Floor, Productivity Press Development Team

Six Sigma

Lean Six Sigma, Michael L. George

Six Sigma for Leadership, Greg Brue, Creative Designs, Inc.

Culture/OLPC

Who Moved My Cheese?: An A-Mazing Way to Deal with Change in Your Work and in Your Life, Spencer Johnson

Zapp!, William C. Byham

Leadership

Execution, Larry Bossidy and Ram Charan

Leadership and the One Minute Manager, Kenneth H. Blanchard, Ph.D.

Lean Enterprise Leader (first 3 chapters), Stephen Hawley Martin

Leading Change, John P. Kotter, Harvard Business School Press

OPTIONAL READING

Manufacturing

Running Today's Factory, Charles Standard and Dale Davis
**America Can Compete!*, James Gooch, Michael George, and Douglas Montgomery
**Manufacturing for Competitive Advantage*, Thomas G. Gunn

Lean—General

Becoming Lean, Jeff K. Liker, Productivity Press
The Lean Office, Collected Practices and Cases originally appeared in the Lean Manufacturing Advisor, 1999–2004. Published by Productivity Press, NY, NY
All I Need to Know About Manufacturing I Learned in Joe's Garage, William B. Miller and Vicki L. Schenk, Bayrock Press
One-Piece Flow, Kenichi Sekine
The Shingo Production Management System—Improving Process Functions, Shigeo Shingo
**Key Strategies for Plant Improvement*, Shigeo Shingo
Lean Solutions, James Womack and Daniel Jones
Henry Ford's Lean Vision, William A. Levinson

Lean—Tools

Integrating Kanban with MRPII, Raymond S. Louis, Productivity Press
A Revolution in Manufacturing: The SMED System, Shigeo Shingo, Productivity Press
Kanban Just-In-Time at Toyota, David J. Lu (Trans.), Productivity Press
The Kaizen Blitz; Accelerating Breakthroughs in Productivity and Performance, Anthony C. Laraia, Patricia E. Moody, and Robert W. Hall, John Wiley and Sons, Inc.
TPM Development Program, Japan Institute for Plant Maintenance, Seiichi Nakajima (Ed.), Productivity Press, Portland, Oregon
Kaizen Strategies for Customer Care, Patricia Wellington
TPM for Every Operator (Shopfloor Series), Japan Institute of Plant Maintenance (Ed.)

Reorganizing the Factory: Competing Through Cellular Manufacturing, Nancy Hyer and Urban Wemmerlov, Productivity Press
Autonomous Maintenance for Operators (Shopfloor Series), Japan Institute of Plant Maintenance (Ed.), Productivity Press
Lean Logistics, Michel Baudin

Leadership

Leading the Lean-Initiative, John W. Davis, Productivity Press
Good to Great, Jim Collins
Kaizen Strategies for Successful Leadership, Tony Barnes
Confronting Reality, Larry Bossidy and Ram Charan
Winning, Jack Welch, Harper Business
Andy & Me: Crisis and Transformation on the Lean Journey, Pascal Dennis, Productivity Press
Why Great Leaders Don't Take Yes for an Answer, Michael A. Roberto
Built to Last, Jim Collins

Culture/OLPC

Strategies for Winning Through People, Sheila Cane
The Heart of Change, John P. Kotter
The Journey to Teams, Michael D. Regan
Winning the Knowledge Transfer Race, Michael English and William Baker, Jr.

General

Leapfrogging the Competition, Oren Harari
The Big Squeeze, Patricia E. Moody
FAST Innovation, Michael George

Appendix D: Formulas for Selected Metrics

SAFETY

OSHA recordable incident rate (ORIR): Number of accidents in specified time period multiplied by 200,000 divided by the total number of hours in the same time period. This is a measure of frequency of occurrence.

Dollars per Hour Worked (DHW): Incurred workers' compensation costs divided by total hours worked. This is a measure of cost.

Reduced Productivity Rate (RPR): [(2) (lost workdays) + restricted workdays] [200,000]/hours worked. This is a measure of accident severity.

QUALITY—INTERNAL FAILURES

Dock Audits (DPMU): The total units rejected from an audit multiplied by 1 million and divided by the total units audited

Internal Rejections (DPMU) = measure of STOP tag only: Determined by the percentage of product that has failed to pass through the entire production process without a defect or rejection (e.g., rework, repair, reinspection) multiplied by 1,000,000. This is the inverse of First Pass Yield.

Gross Waste %: Value of all direct materials, labor, and overhead of scrapped material divided by the total cost of production

QUALITY—EXTERNAL METRICS

Customer Complaints: Track the number of complaints. Report the total complaints received per month broken down into the specific reasons for the complaint, e.g. incorrect price on the invoice, freight damage, quality issue, late delivery, etc.

Number of Returns: Measure the number of complaints that result in an approval for the customer to return the product for credit, rework or replacement.

Allowance/Adjustments: Measure the costs that are lost due to complaints. This could be the amount of price adjustments for subnormal quality, credit for the amount of product that could not be used and was scrapped, the cost of freight to return the material, etc.

SERVICE

Workorder Performance to current schedule: The measurement is expressed as the percentage of workorders completed on the day or within the week of its current schedule date.

$$\frac{\text{Workorders Completed On Time}}{\text{Total Workorders Due Current Period}}$$

The measurement includes released workorders for finished goods or subassemblies and raw materials tied only to an internal or external customer order. There is no tolerance on the due date such that early deliveries and late deliveries are both "a miss." The ordered quantity must be produced plus or minus 5%* to be considered complete.

Workorder Performance to original schedule: The measurement is expressed as the percentage of workorders completed when it was originally scheduled, that is, initial due date assigned to the workorder.

* These numbers should, of course, be adjusted based on the reader's business requirements.

$$\frac{\text{Workorders Completed On Time to Original Schedule}}{\text{Total Original Schedule Workorders Due}}$$

Workorders completed on time to original schedule. Total original schedule workorders due. The measurement includes released workorders for finished goods or subassemblies and raw materials tied to only an internal or external customer order. There is no tolerance on the dates such that early deliveries and late deliveries are both "a miss." The ordered quantity must be produced plus or minus 5% to be considered complete.

EFFECTIVENESS

Overall Equipment Effectiveness: The product of utilization and efficiency and first-pass yield (failing that breakout of information, it is actual standard hours earned divided by total scheduled hours for DL associates). I strongly recommend measuring all 3 components.

Breakdown Maintenance Rate: Unscheduled maintenance hours/total maintenance hours. Unscheduled means it was not planned to be done at the time there was a need for maintenance.

Cost of Maintenance: The sum of maintenance labor dollars and maintenance materials dollars divided by the plant labor and overhead costs for the month

WORKING CAPITAL MANAGEMENT

RIP Turns (RIP is the sum of raw material and work-in-process inventories): Three-month inventory transfers to finished goods annualized divided by three-month average RIP inventory. This measures velocity of the flow through the plant.

Cycle Count Accuracy: A measurement, expressed as a percentage of the accuracy of the inventory system. The percentage is based on, for example, errors found on quantities, locations, and identification of materials versus the number of opportunities. It is not based on dollars of accounting adjustments (though finance managers may want that tracking in addition).

Appendix E: Example of a Communications Plan Calendar 20XX

Disclaimer and terms of use: The Manufacturing Excellence Audit (MEA), Example of a Communications Plan, and Example of a Training Plan are Excel workbooks with standard working formulas. Their purpose is to provide the user with sample templates for reference. User agrees to release author and support personnel from any and all liability associated with the use of the formulas used in these workbooks.

Refer to Appendix E on the CD-Rom inside the back cover.

Appendix F: Example of a Training Plan

Disclaimer and terms of use: The Manufacturing Excellence Audit (MEA), Example of a Communications Plan, and Example of a Training Plan are Excel workbooks with standard working formulas. Their purpose is to provide the user with sample templates for reference. User agrees to release author and support personnel from any and all liability associated with the use of the formulas used in these workbooks.

Refer to Appendix F on the CD-Rom inside the back cover.

Index

D

H

I

J

K

L

M

Q